On the Safety Lamp for Preventing Explosions in Mines,

Houses Lighted by Gas, Spirit Warehouses, or Magazines in Ships, &c

HUMPHRY DAVY

CAMBRIDGE
UNIVERSITY PRESS

CAMBRIDGE UNIVERSITY PRESS

Cambridge, New York, Melbourne, Madrid, Cape Town,
Singapore, São Paolo, Delhi, Mexico City

Published in the United States of America by Cambridge University Press, New York

www.cambridge.org
Information on this title: www.cambridge.org/9781108052139

© in this compilation Cambridge University Press 2012

This edition first published 1825
This digitally printed version 2012

ISBN 978-1-108-05213-9 Paperback

CAMBRIDGE LIBRARY COLLECTION

Books of enduring scholarly value

Technology

The focus of this series is engineering, broadly construed. It covers technological innovation from a range of periods and cultures, but centres on the
technological achievements of the industrial era in the West, particularly
in the nineteenth century, as understood by their contemporaries. Infra-
structure is one major focus, covering the building of railways and canals,
bridges and tunnels, land drainage, the laying of submarine cables, and
the construction of docks and lighthouses. Other key topics include
developments in industrial and manufacturing fields such as mining
technology, the production of iron and steel, the use of steam power, and
chemical processes such as photography and textile dyes.

On the Safety Lamp for Preventing Explosions in Mines, Houses Lighted by Gas, Spirit Warehouses, or Magazines in Ships, &c

Self-taught chemist and inventor Sir Humphry Davy (1778–1829) was one
of the first professional scientists of his age. President of the Royal Society
from 1820 to 1827, he was also a brilliant lecturer whose popularising of
science made him famous. He also pioneered electrochemistry, isolating
potassium, sodium and calcium. But Davy is best known for creating
the safety lamp when he was asked to address the frequent occurrence of
explosions in coal mines. He realised that firedamp – flammable gases such
as methane – was ignited at high temperature by the open flames of miners'
lamps. In 1815, he devised a lamp with a mesh screen that prevented
ignition of firedamp; this application of science allowed miners to work in
greater safety. First published in 1818 and revised in 1825, this work details
the invention that cemented Davy's position as a national hero and earned
him the Royal Society's Rumford Medal.

Cambridge University Press has long been a pioneer in the reissuing of out-of-print titles from its own backlist, producing digital reprints of books that are still sought after by scholars and students but could not be reprinted economically using traditional technology. The Cambridge Library Collection extends this activity to a wider range of books which are still of importance to researchers and professionals, either for the source material they contain, or as landmarks in the history of their academic discipline.

Drawing from the world-renowned collections in the Cambridge University Library and other partner libraries, and guided by the advice of experts in each subject area, Cambridge University Press is using state-of-the-art scanning machines in its own Printing House to capture the content of each book selected for inclusion. The files are processed to give a consistently clear, crisp image, and the books finished to the high quality standard for which the Press is recognised around the world. The latest print-on-demand technology ensures that the books will remain available indefinitely, and that orders for single or multiple copies can quickly be supplied.

The Cambridge Library Collection brings back to life books of enduring scholarly value (including out-of-copyright works originally issued by other publishers) across a wide range of disciplines in the humanities and social sciences and in science and technology.

ON THE SAFETY LAMP

FOR

COAL MINERS,

WITH SOME

RESEARCHES ON FLAME.

ON THE

Safety Lamp

FOR

PREVENTING EXPLOSIONS IN MINES,

HOUSES LIGHTED BY GAS,

SPIRIT WAREHOUSES,

OR

MAGAZINES IN SHIPS, &c.

WITH SOME

RESEARCHES ON FLAME.

By SIR HUMPHRY DAVY, Bart.

PRESIDENT OF THE ROYAL SOCIETY, &c.

LONDON:

PRINTED FOR R. HUNTER,

(SUCCESSOR TO MR. JOHNSON,)

No. 72, ST. PAUL'S CHURCH-YARD.

1825.

H. Bryer, Printer, Bridge-street, Blackfriars, London.

ADVERTISEMENT.

THIS work was published in 1818, but a part of the edition having remained unsold, and the recent occurrence of some severe accidents from explosions in coal mines, and houses lighted by gas, and the loss of ships by fire, having shewn, that the precautions which it was intended to describe, either are not known or are not attended to, I have thought it might assist the cause of humanity, to advertise the book a second time. I have added to it a few additional paragraphs, which contain some new facts and some practical results, connected with the use of the Safety Lamp; most of the last occurred iv.

to me during journeys that I have made, for the purpose of introducing this invention into the principal mines of Europe, in which inflammable air is found.

Park Street,
March, 20, 1825.

PREFACE.

———

I HAVE thought it right to collect and to publish in a connected form an account of all the researches that I have made on the subject of explosions from inflammable air, and the modes in which they may be prevented, as well as the collateral investigations to which they have given rise, with the hope of presenting a permanent record on this important subject to the practical miner, and of enabling the friends of humanity to estimate and apply those resources of science, by which a great and permanently existing evil may be subdued.

In connecting general views on the subject with extracts from papers published in the Transactions of the Royal Society, I fear I shall sometimes be accused of repetitions; but in a case where human life is concerned, and by which human happiness may be deeply affected, I shall not dread the accusation of dwelling too long upon, or treating too often of, precautionary measures.

The names of three persons will be found mentioned in these pages as having assisted in the investigations. The public owe much to the Rev. John Hodgson and to Mr. Buddle for having been the first persons to make experiments upon the Safety Lamp in explosive atmospheres in the mine, and for elucidating its practical application and rendering it familiar to the miner ; and I am myself indebted to Mr. Michael Faraday for much able assistance in the prosecution of my experiments,

I have given the extracts from my papers
nearly in the order in which they were
published; which will, I hope, both render
the facts more intelligible, and shew the
gradual progress of the enquiry: in which
every step was furnished by experiment or
induction, in which nothing can be said to
be owing to accident, and in which the
most simple and useful combination arose
out of the most complicated circumstances.

The result of these labours will, I trust,
be useful to the cause of science, by proving that even the most apparently abstract
philosophical truths may be connected with
applications to the common wants and purposes of life.

The gratification of the love of knowledge
is delightful to every refined mind; but a
much higher motive is offered for indulging
in it, when that knowledge is felt to be

practical power, and when that power may be applied to lessen the miseries or increase the comforts of our fellow creatures.

London, May 14, 1818.

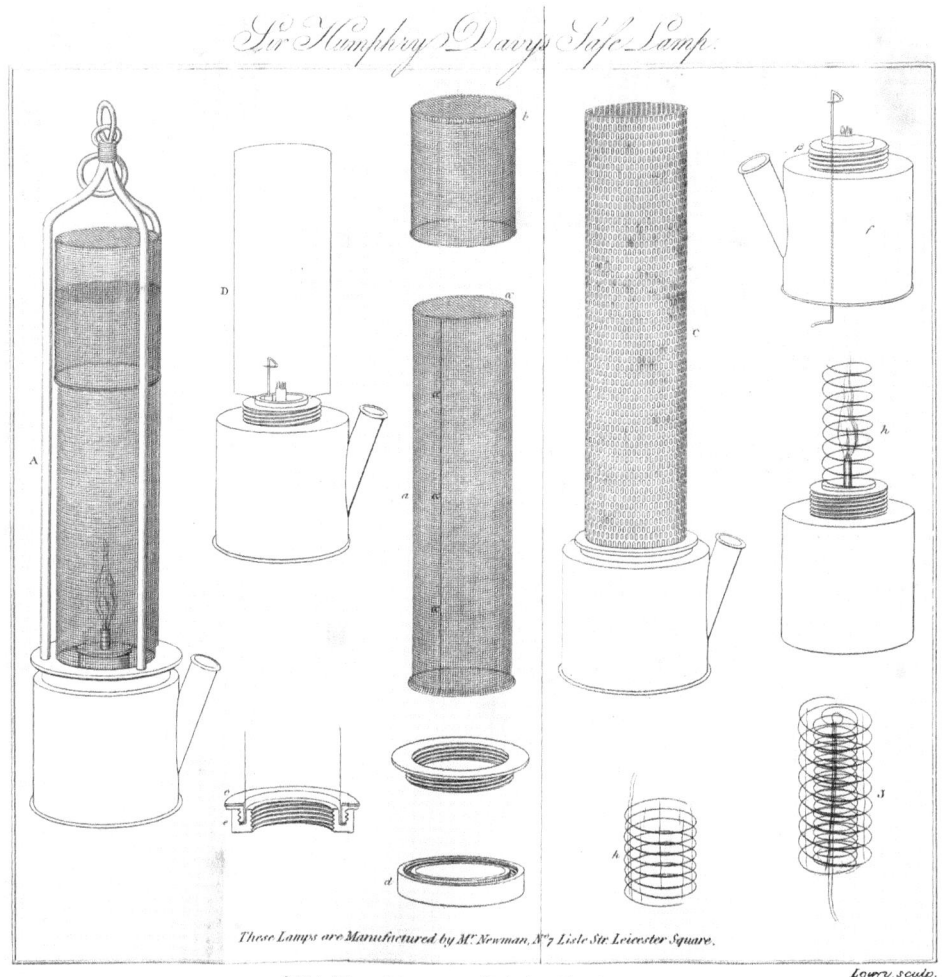

Sir Humphry Davy's Safe Lamp.

These Lamps are Manufactured by Mr Newman, Nº 7 Lisle Str Leicester Square.

Published May, 23, 1818. by R. Hunter, St Pauls Church Yd London.

Lowry, sculp.

The material originally positioned here is too large for reproduction in this reissue. A PDF can be downloaded from the web address given on page iv of this book, by clicking on 'Resources Available'.

THE SAFETY LAMP;

WITH

RESEARCHES ON FLAME.

I. *General Views of the Progress of the Researches on the Safety Lamp, and of the Principles on which its Security depends.*

THE use of pit coal in Britain is connected not only with the necessities, comforts, and enjoyments of life; but also with the extension of our most important arts, our manufactures, commerce, and national riches. Essential in affording warmth and preparing food, it yields a sort of artificial sunshine, and in some measure compensates for the disadvantages of our climate. —By means of it, metallurgical processes are carried on, and the most important materials of civilized life furnished, the agriculturist is supplied with an useful manure, and the architect with a necessary cement. Not only manufactories and private houses,

B

but even whole streets are lighted by its application—and in furnishing the element of activity in the steam engine, it has given a wonderful impulse to mechanical and chemical ingenuity, diminished to a great extent human labour, and increased in a high degree the strength and wealth of the country.

Every thing connected with the permanent supply of such a material, is worthy of scientific consideration ; and to remove obstacles, difficulties, or dangers connected with its production, is not unimportant to the state.

Since the earliest period of the application of mineral coal* to the purposes of fuel, the explosions in coal mines from inflammable air† or fire damp, have been regarded as the greatest evil occurring in the working of the mines. The strata of coal lie usually parallel or nearly parallel to the surface, at certain depths beneath it, and

* Coal was certainly worked in the neighbourhood of Newcastle in 1245. See Brande's History of Newcastle, Vol. II. p. 253.

† Called Grisoux in Flanders, and Feu Brisson in the southern departments of France.

this coal, when the pressure of the super-incumbent material is removed, affords inflammable air; which is disengaged not only in the common operations of mining, when the coal is broken and removed, but is likewise permanently evolved, often in enormous quantities,* from fissures in the strata.

When it has accumulated in any part of the gallery or chamber of a mine, so as to be mixed in certain proportions with common air, the presence of a lighted candle or lamp causes it to explode, and to destroy, injure, or burn whatever is exposed to its violence.

To give detailed accounts of the tremendous accidents, owing to this cause, would be merely to multiply pictures of death and of human misery.—The phenomena are always of the same kind.—The miners are either immediately destroyed by the explosion, and thrown with the horses and machinery through the shaft into the air, the mine becoming as it were an enormous

---

* This phenomenon is called, in the language of the North country miners, a *blower*.

piece of artillery, from which they are pro-
jected ; or they are gradually suffocated,
and undergo a more painful death from the
carbonic acid and azote remaining in the
mine after the inflammation of the fire
damp; or what, though it appears the
mildest, is perhaps the most severe fate,
they are burnt or maimed, and often rendered incapable of labour and of healthy
enjoyment for life.

The fire damp is found in the greatest
quantity, and is most dangerous, in the
deepest mines; but it likewise often occurs
in superficial excavations, and I have now
a letter, of the date of June 8, 1816, in
my possession, in which it is stated, that
in the very commencement of working a
coal mine in Shropshire, several miners
were killed, and others severely burnt.

Modes of preventing accidents from fire
damp have been ardently sought for by all
persons connected with coal mines, and it
has even occupied the attention of an en-
lightened government.—In consequence of
some explosions which prevented the miners
from working the coal mines at Briançon,

in Dauphiny,* the Duke de Choiseul, at that time Prime Minister of France, recommended the subject to the consideration of the Academy of Sciences, and a committee was appointed, who made it for some time the object of their attention; but the plan that they proposed for avoiding the danger, was a common mode of ventilation.

This evil of the fire damp, though belonging to all coal mines, has been most severely experienced in those of Hainault, in Flanders, and the infinitely more important mines in the neighbourhood of Whitehaven and Newcastle, in this country.

The number of dreadful accidents, indeed, which had happened within the last three or four years in the last mentioned districts, particularly that by which ninety-six persons were destroyed in the Felling colliery, had so strongly impressed the minds of a number of benevolent persons belonging to or connected with the coal districts, that it was said to be in their contemplation to bring the subject before Parliament, that by making it a national question, it

* Histoire de l'Academie Royale, 1763, p. 1.

might obtain that consideration which its importance demanded.

When I first turned my attention particularly to the subject, which was in August, 1815, in consequence of a letter from the Rev. Dr. Gray, there appeared very little hope of finding an efficacious remedy. The resources of modern chemical science had been fully applied in ventilation, in the improved plans of Mr. Buddle; the comparative lightness of the fire damp was well understood, every precaution was taken to preserve the communications open; and the currents of air were promoted or occasioned, not only by furnaces, but likewise by air pumps and steam apparatus.

Sir James Lowther had observed early in the last century that the fire damp in its usual form was not inflammable by sparks from flint and steel; and a person in his employment had invented a mill for giving light by the collision of flint and steel,* and this was the only instrument except common candles employed in the

* Said to be Mr. Spedding, in Hutchinson's History of Cumberland. Article, Whitehaven.

dangerous parts of the British collieries. Yet instances of explosion have been known from the steel mill, and it required manual labour for its use. In Flanders amadou or fungus tinder had been occasionally employed in dangerous parts of the mine, but the light yielded by this substance was much too feeble to be used for working the mines, and only enabled the miners to find their way for particular occasions.

M. de Humboldt, the justly celebrated philosophical traveller, in 1796 conceived and executed the plan of a lamp* for giving light in mines where a common candle would not burn or produce explosion; but it was founded on the principle of entire insulation from the air, and could burn only for a short time till the air contained within it was exhausted. A lamp upon a plan similar as to insulation, was contrived by Dr. Clanny, in 1813, but he supplied his light with air from the mine through water by bellows, and it went out in explosive atmospheres, and to be employed required to be worked by hand, or by machinery; and neither

* Journal des Mines. Tom. VIII. p. 839.

M. de Humboldt's lamps, or Dr. Clanny's, had, for obvious reasons I believe, ever been used in coal mining.

The great object, one rather to be ardently desired than confidently expected, was to find a light, which at the same time that it enabled the miner to work with security in explosive atmospheres, should likewise consume the fire damp. Having learnt from Mr. Buddle the degree of light required for the common operations of the workmen, I made several experiments with the hope of obtaining such a light without active inflammation. I tried Kunckel's, Canton's, and Baldwin's phosphorus, and likewise the electrical light in close vessels, but without success. I had a lamp made with two valves, which closed in atmospheres contaminated with fire damp, by the increased heat of the flame produced by the combustion of the gas, but this lamp could not be used in an explosive atmosphere.

It will be unnecessary to dwell upon preliminary and unsuccessful attempts, and I shall proceed to describe the origin and progress of those investigations which led

me to the discovery of the principles by
which explosion and flame may be arrested
and regulated; and by means of which, the
miner is not only able to subdue and con-
troul, but likewise to render useful his most
dangerous enemy.

I first began with a minute chemical examination of the substance with which I
had to contend.—The analysis of various
specimens of fire damp shewed me that the
pure inflammable part of it was light carburetted hydrogen, as Dr. Henry had before
stated, hydrogen or pure inflammable air
combined with charcoal or carbon.

I made numerous experiments on the circumstances under which it explodes, and the
degree of its inflammability.—I found that
it required to be mixed with very large
quantities of atmospheric air to produce explosion, even when mixed with three or
nearly four times its bulk of air, it burnt
quietly in the atmosphere, and extinguished
a taper. When mixed with between five and
six times its volume of air it exploded feebly:
it exploded with most energy when mixed
with seven or eight times its volume of air,
and mixtures of fire damp and air retained

their explosive power when the proportions were one of gas to fourteen of air. When the air was in larger quantity the flame of a taper was merely enlarged in the mixture, an effect which was still perceived in thirty parts of air to one of gas.

I found the fire damp much less combustible than other inflammable gases. It was not exploded or fired by red hot charcoal or red hot iron; it required iron to be white hot, and itself in brilliant combustion, for its inflammation. The heat produced by it in combustion was likewise much less than that of most other inflammable gases, and hence, in its explosion, there was much less comparative expansion.

On mixing 1 part of carbonic acid or fixed air with 7 parts of an explosive mixture of fire damp, or 1 part of azote with 6 parts, their powers of exploding were destroyed.

In exploding a mixture in a glass tube of one-fourth of an inch in diameter and a foot long, more than a second was required before the flame reached from one end to the other: and I found that in tubes of one-seventh of an inch in diameter, explosive mixtures could not be fired when they were opened in the atmosphere; and that metallic tubes prevented explosion better than glass tubes.

In reasoning upon these various phenomena it occurred to me, as a *considerable* heat was required for the inflammation of the fire damp, and as it produced in burning comparatively a *small degree* of heat, that the effect of carbonic acid and azote, and of the surfaces of small tubes in preventing its explosion, depended upon their cooling powers; upon their lowering the temperature of the exploding mixture so much that it was no longer sufficient for its continuous inflammation.

This idea, which was confirmed by various obvious considerations, led to an immediate result—the possibility of constructing a lamp, in which the cooling powers of the azote or carbonic acid, formed by combustion or the cooling powers of the apertures, through which the air entered or made its exit, should prevent the communication of explosion.

I first tried the effects of lamps in which there was a very limited circulation of air; and I found that when a taper in a close lantern was supplied with air so as to burn feebly from very small apertures below the flame; and at a considerable distance from it, it became extinguished in explosive mixtures; but I ascertained that precautions which it would be dangerous to trust to workmen were required to make this form of a lamp safe, and that at best it could give only a feeble light; and I immediately adopted systems of tubes above and below, of that diameter in which I had ascertained that explosions would not take place.

In trying my first tube lamp in an explosive mixture I found that it was safe; but unless the tubes were very short and numerous, the flame could not be well supported; and in trying tubes of the diameter of one-seventh or one-eighth of an inch I determined that they were safe only to small quantities of explosive mixture, and when of a given length; and that tubes even of a much smaller diameter communicated explosion from a close vessel. Hence I took a new method of ascertaining the safety of my apertures, and of trying different forms of apertures.

I had a vessel furnished with wires by

which the electrical spark could be taken
in an explosive mixture, and which was
larger in capacity than a safe lamp or
lantern was required to be. I placed my
flame sieves, i. e. my systems of apertures,
between this jar and a bladder containing
likewise an explosive mixture, and I judged
the apertures to be safe only when they
stopped explosion acting upon them in this
concentrated way.

In this mode of experimenting I soon
discovered that a *few apertures* even of very
small diameter were not safe unless their
sides were very deep; that a single tube of
one-twenty-eighth of an inch in diameter
and two inches long suffered the explosion
to pass through it; and that a *great number*
of small tubes, or of apertures, stopped explosion even when the depths of their sides
was only equal to their diameters—and at
last I arrived at the conclusion that a *metallic tissue*, however thin and fine, of which
the apertures filled more space than the
cooling surface, so as to be permeable to air
and light, offered a perfect barrier to explosion, from the force being divided be-

tween, and the heat communicated to, an immense number of surfaces.

My first safety lamps constructed on these principles, gave light in explosive mixtures containing a great excess of air, but became extinguished in explosive mixtures in which the fire damp was in sufficient quantity to absorb the whole of the oxygen of the air, so that such mixtures never burnt continuously at the air feeders, which in lamps of this construction was important, as the increase of heat, where there was only a small cooling surface, would have altered the conditions of security.

I made several attempts to construct safety lamps which should give light in all explosive mixtures of fire damp, and after complicated combinations I at length arrived at one evidently the most simple, that of surrounding the light entirely by wire gauze, and making the same tissue feed the flame with air and emit light.

In plunging a light surrounded by a cylinder of fine wire gauze into an explosive mixture I saw the whole cylinder become quietly and gradually filled with flame, the

upper part of it soon appeared red hot; yet
no explosion was produced.

It was easy at once to see that by increasing the cooling surface in the top, or in any
other part of the lamp, the heat acquired by
it might be diminished to any extent; and
I immediately made a number of experiments to perfect this *invention* which was
evidently the one to be adopted, as it ex-
cluded the necessity of using glass or any
fusible or brittle substance in the lamp, and
not only deprived the fire damp of its explosive powers, but rendered it an useful
light.

Though all the specimens of fire damp
which I had examined consisted of carburetted hydrogen mixed with different small
proportions of carbonic acid and common
air, yet some phenomena that I observed in
the combustion of a blower, induced me to
believe that small quantities of olefiant gas
may be sometimes evolved in coal mines
with the carburetted hydrogen.—I therefore
resolved to make all lamps safe to the test
of the gas produced by the distillation of
coal, which when it has not been exposed
to water always contains olefiant gas.—

1 placed my lighted lamps in a large glass receiver, through which there was a current of atmospherical air, and by means of a gasometer filled with coal gas, I made the current of air which passed into the lamp more or less explosive, and caused it to change rapidly or slowly at pleasure, so as to produce all possible varieties of inflammable and explosive mixtures; and I found that iron wire gauze, composed of wires from one-fortieth to one-sixtieth of an inch in diameter, and containing twenty-eight wires or 784 apertures to the inch, was safe under all circumstances in atmospheres of this kind: and I consequently adopted this material in guarding lamps for the coal mines, where in January, 1816, they were immediately adopted, and have long been in general use.

Observations upon them in their working state, and upon the circumstances to which they are exposed, have led to a few improvements or alterations, merely connected with the modes of increasing light or diminishing heat, which were very obvious from the original construction; and experiments on the nature of flame, and the modifications of combustion led me, in January, 1817, to an important practical addition, founded entirely upon a new principle.

The wire gauze lamp in its common form burns in all atmospheres that are explosive, and by suspending or placing in it a little cage of wire of platinum or palladium from one-sixtieth to one-seventieth of .an inch in thickness, it yields a light in atmospheres too much contaminated with fire damp to be explosive, a slow combination being occasioned by the heated platinum between the elements of the gas and oxygen which produces sufficient heat to keep the metals of low conducting power and low capacity for heat permanently ignited wherever there is air enough to support life without suffering.

I shall conclude this view by some general observations on flame and combustion which will shew more distinctly the causes and the limits of safety in lamps, and which will demonstrate the danger of combinations made with an imperfect knowledge of the principles of security.

Flame may be defined to be aeriform or gaseous matter, heated to such a degree as to be luminous, and it may be produced

c

independent of any chemical changes, as is shewn in the discharge of voltaic electricity through an undecomposable gas. Very concentrated electricity, in passing through bodies, constantly heats them whether they be solid, fluid, or gaseous : and by the voltaic apparatus, the nature of flame is distinctly shewn. Flames are conical because the greatest heat is in the centre of the mass, and because heated air rapidly ascends through cooler air. When *flame* is produced in chemical combination, gaseous matter is the cause of it ; and the heat of flames seems always proportional, other circumstances remaining the same to the rapidity of combination, and to the density of the gases combining. Thus the heat of flames diminishes by rarefaction, and increases by condensation. Whenever combustible gaseous matter burns in the atmosphere, it must first mix with a certain quantity of air : if it require a high temperature for its combustion, it will be easily extinguished by rarefaction, or by cooling agencies, whether of solid surfaces, or mixtures of incombustible gases : if it require a very low temperature for its combustion, it will burn in highly rarefied air, or under considerable cooling agencies.

By heating strongly gases that burn with difficulty, their continued inflammation becomes easy, in consequence of increments of heat occasioned by combustion of small quantities, which under any other circumstances would not produce continued combustion. Hence if mixtures of fire damp are burnt from systems of tubes or canals, or metallic plates, which have small radiating and cooling surfaces : though these systems are safe at first, they become dangerous as they are heated.* Where currents are occasioned which concentrate explosive mixtures by the air feeders in lamps being below,* and made in thick metallic plates or canals; there being an increment of heat within, and a very small radiating surface without, as the heat increases, the combustion of the explosive mixture will gradually extend further, and at last com-

* I warn the coal miner against any pretended safety lamps made in this manner, and which, to superficial observers, may appear to be constructed upon principles of security, but in which these principles cannot really exist.

municate with the external air; for explosion will be communicated by any aperture, however small, provided it be sufficiently heated. This circumstance is shewn
in a very elegant manner in burning concentrated mixtures of oxygen and hydrogen
at the end of a long tube of one-sixtieth or
one-seventieth of an inch in diameter, when
the experiment begins, (the tube being cool)
there is no danger; gradually, however, as
it becomes heated, the combustion steals as
it were down the tube, and at last reaches
the reservoir of the gases.

Where one set of air feeders only are at-
tempted in a lamp, they should present an
uniform surface, so that the radiating powers
of the metal, and the cooling powers of the
external gases, may immediately balance
the heating powers of the internal gases;
and where the radiating tissue is connected
with other parts of the lamp, these parts
should be so massy as to be slightly heated
only, and present no means for a gradual
accumulation of heat. Wire gauze, as it
offers a greater extent of radiating surface
than perforated metallic plates, is the best
material for the guard of lamps; and by

being made of the proper degree of fineness, it will form a barrier for every species of explosion requiring temperatures higher than those of our atmosphere : but the apertures must be smaller, and the radiating surfaces greater, in proportion to the inflammability of the gas : and currents of concentrated explosive mixtures, acting even for any length of time, may be stopped by reduplications of wire gauze. Wire gauze for lamps must not be made of, or covered with any easily combustible metal : fine brass wire is improper, on account of the zinc it contains, and the iron wire should not be tinned. The body of the lamp should be of copper riveted together, or of massy cast brass or cast iron ; the screws should fit tight ; no aperture, *however small*, should be suffered to exist in the body of the lamp, and the trimming wire should move through a long tight tube.

Flame, whether produced by the combustion of large or small quantities of explosive mixture, may be always extinguished or destroyed by certain cooling agencies ; and in proportion to the heat required to carry on the combustion, so is it more easily destroyed. The temperature of metal even when white hot, is far below that of flame; and hence red hot gauze, in sufficient quantity, and of the proper degree of fineness, will abstract sufficient heat from the flame of carburetted hydrogen or fire damp, to extinguish it.

Combinations of gases may be occasioned by a heat not sufficient to raise their temperature into flame, but they still produce heat during their combination, as is evident from what has been stated page 17, and when in a mixture containing air and combustible gases, the cooling agencies are too great to permit the appearance of flame or their continued combination; still this combination may be kept up by the ignition of platinum, so that with a certain quantity of platinum in a cage of wire gauze, the fire damp may be entirely consumed without flame, yielding only a beautiful light by the ignition of solid matter.

II. *Extracts from Papers published in the Philosophical Transactions, and in the Journal of Science and the Arts, on the Fire Damp, the Safety Lamp, and on Flame.*

---

1. *On the Fire Damp of Coal Mines, and on Methods of lighting the Mines, so as to prevent its Explosion.* Nov. 9, 1815.*

" THE fire damp is produced in small quantities in coal mines, during the common process of working.

The Rev. Mr. HODGSON informed me, that on pounding some common Newcastle coal fresh from the mine in a cask furnished with a small aperture, the gas from the aperture was inflammable. And on breaking some large lumps of coal under water, I ascertained that they gave off inflammable gas†. Gas is likewise disen-

---

* From PHIL. TRANS.

† This is probably owing to the coal strata having been formed under a pressure greater than that of the

gaged from bituminous shist, when it is worked.

The great sources of the fire-damp in mines are, however, what are called blowers, or fissures in the broken strata, near dykes, from which currents of fire-damp issue in considerable quantity, and sometimes for a

atmosphere, so that they give off elastic fluid when they are exposed to the free atmosphere : and probably coals containing animal remains, evolve not only the fire-damp, but likewise azote and carbonic acid.

In the Apennines, near Pietra Mala, I examined a fire produced by gaseous matter, constantly disengaged from a shist stratum : and from the results of the combustion, I have no doubt but that it was pure fire-damp. Mr. M. Faraday, who accompanied me, and assisted me in my chemical experiments, in my journey, collected some gas from a cavity in the earth about a mile from Pietra Mala, then filled with water, and which, from the quantity of gas disengaged, is called Aqua Buja. I analysed it in the Grand Duke's laboratory at Florence, and found that it was pure light carburetted hydrogene, requiring two volumes of oxygene for its combustion, and producing a volume of carbonic acid gas.

It is very probable, that these gases are disengaged from coal strata beneath the surface, or from bituminous shist above coal ; and at some future period new sources of riches may be opened to Tuscany from this invaluable mineral treasure, the use of which in this country has supplied such extraordinary resources to industry.

long course of years.*　When old workings are broken into, likewise, they are often found filled with fire-damp; and the deeper the mine the more common in general is this substance.

I have analysed several specimens of the fire-damp in the laboratory of the Royal Institution; the pure inflammable part was the same in all of them, but it was sometimes mixed with small quantities of atmospherical air, and in some instances with azote and carbonic acid.

Of six specimens collected by Mr. Dunn from a blower in the Hepburn Colliery, by emptying bottles of water close to it, the purest contained one-fifteenth only of atmospherical air, with no other contamination, and the most impure contained five-twelfths of atmospherical air; so that this air was probably derived from the circumambient air of the mine. The weight of the purest specimen was for 100 cubical inches 19.5 grains.

One measure of it required for its com-

* Sir James Lowther found a uniform current produced in one of his mines for two years and nine months. Phil. Trans. Vol. XXXVIII. p. 112.

plete combustion by the electric spark nearly two measures of oxygene, and they formed nearly one measure of carbonic acid.

Sulphur heated strongly and repeatedly sublimed in a portion of it freed from oxygene by phosphorus, produced a considerable enlargement of its volume, sulphuretted hydrogene was formed, and charcoal precipitated; and it was found that the volume of the sulphuretted hydrogene produced, when it was absorbed by solution of potassa, was exactly double that of the fire-damp decomposed.

It did not act upon chlorine in the cold; but, when an electric spark was passed through a mixture of 1 part of it with 2 of chlorine, there was an explosion, with a diminution to less than one-fourth, and much charcoal was deposited.

The analysis of specimens of gas sent to my friend JOHN GEORGE CHILDREN, Esq. by Dr. CLANNY, afforded me similar results; but they contained variable quantities of carbonic acid gas and azote.

Different specimens of these gases were tried by the test of exposure to chlorine both in darkness and light: they exhibited no marks of the presence of olefiant gas or hydrogene; and the residuum produced by detonation with chlorine showed them to be free from carbonic oxide.

It is evident, then, that the opinion formed by other chemists respecting the fire-damp is perfectly correct; and that it is the same substance as the inflammable gas of marshes, the exact chemical nature of which was first demonstrated by Mr. DALTON; and that it consists, according to my view of definite proportions, of 4 proportions of hydrogene in weight 4, and 1 proportion of charcoal in weight 11.5.

I made several experiments on the combustibility and explosive nature of the fire-damp. When 1 part of fire-damp was mixed with 1 of air, they burnt by the approach of a lighted taper, but did not explode; 2 of air and 3 of air to 1 of gas produced similar results. When 4 of air and 1 of gas were exposed to a lighted candle, the mixture being in the quantity of 6 or 7 cubical inches in a narrow necked bottle, a flame descended through the mixture, but there was no noise: 1 part of gas inflamed with 6 parts of air in a similar bottle, produced a slight whistling sound: 1 part of gas with 8 parts of air, rather a louder sound: 1 part with 10, 11, 12, 13 and 14 parts, still inflamed, but the violence of combustion diminished. In 1 part of gas and 15 parts of air, the candle burnt without explosion with a greatly enlarged flame; and the effect of enlarging the flame, but in a gradually diminishing ratio, was produced as far as 30 parts of air to 1 of gas.

The mixture which seemed to possess the greatest explosive power, was that of 7 or 8 parts of air to 1 of gas; but the report produced by 50 cubical inches of this mixture was less than that produced by one-tenth of the quantity of a mixture of 2 parts of atmospherical air and 1 of hydrogene.

It was very important to ascertain the degree of heat required to explode the fire-damp mixed with its proper proportion of air.

I found that a common electrical spark would not explode 5 parts of air and 1 of fire-damp, though it exploded 6 parts of air and 1 of damp: but very strong sparks from the discharge of a Leyden jar, seemed to have the same power of exploding different mixtures of the gas as the flame of the taper. Well burned charcoal, ignited to the strongest red heat, did not explode any mixture of air and of the fire-damp; and a fire made of well burned charcoal, i. e. charcoal that burned without flame, was blown up to whiteness by an explosive mixture containing the fire-damp, without producing its inflammation. An iron rod at the highest degree of red heat, and at the common degree of white heat, did not inflame explosive mixtures of the fire-damp; but, when in brilliant combustion, it produced the effect.

The flame of gaseous oxide of carbon as well as of olefiant gas exploded the mixtures of the fire-damp.

In respect of combustibility, then, the fire-damp differs most materially from the other common inflammable gases. Olefiant gas, which I have found explodes mixed in the same proportion with air, is fired by both charcoal and iron heated to redness. Gaseous oxide of carbon, which explodes when mixed with 2 parts of air, is likewise inflammable by red hot iron and charcoal. And hydrogene, which explodes when mixed with three-sevenths of its volume of air, takes fire at the lowest visible heat of iron and charcoal; and the case is the same with sulphuretted hydrogene.

I found that when 6 of air and 1 of fire-damp were exploded over water by a strong electrical spark, the explosion was not very strong, and, at the moment of the greatest expansion, the volume of the gas did not appear to be increased more than one-half*.

I mixed azote and carbonic acid in different quantities with explosive mixtures of fire-damp, and I found that even in very small proportions they diminished the velocity of the inflammation. Azote, when mixed in the proportion of 1 to 6 of an explosive mixture, containing 12 of air and 1 of fire-damp, deprived it of its power of explosion; when 1 part of azote was mixed with 7 of an explosive mixture, only a feeble blue flame slowly passed through the mixture.

1 part of carbonic acid to 7 of an explosive mixture deprived it of the power of exploding; so that its effects are more re-

---

* This appears the expansion when the tube is very small; in larger tubes, it is considerably more. The volume of the gas appears at least tripled during the explosion.

markable than those of azote; probably, in consequence of its greater capacity for heat, and probably, likewise, of a higher conducting power connected with its greater density.

In exploding a mixture of 1 part of gas from the distillation of coal, and 8 parts of air in a tube of a quarter of an inch in diameter and a foot long, more than a second was required before the flame reached from one end of the tube to the other; and I could not make any mixture explode in a glass tube one-seventh of an inch in diameter; and this gas was more inflammable than the fire-damp, as it consisted of carburetted hydrogene gas mixed with some olefiant gas.

In exploding mixtures of fire-damp and air in a jar connected with the atmosphere by an aperture of half an inch, and connected with a bladder by a stopcock, having an aperture of about one-sixth of an inch,* I found

* Since these experiments were made, Dr. Wol- laston has informed me, that Mr. Tennant had observed some time ago, that mixtures of the gas from the distillation of coal and air, would not explode in very small tubes.

that the flame passed into the atmosphere but did not communicate through the stop-cock, so as to inflame the mixture in the bladder: and in comparing the power of tubes of metal and those of glass, it appeared that the flame passed more readily through glass tubes of the same diameter; and this phenomenon probably depends upon the heat lost during the explosion in contact with so great a cooling surface, which brings the temperature of the first portions exploded below that required for the firing of the other portions. Metal is a better conductor of heat than glass: and it has been already shown that the fire-damp requires a very strong heat for its inflammation."

I found, likewise, that inflammation could not be communicated to explosive mixtures in long narrow canals of metal, or in short canals of smaller diameter, or through fine metallic gauze of the kind used for sieves.

" The consideration of these various facts, led me to adopt a form of a lamp, in which the flame, by being supplied with only a limited quantity of air, should produce such a quantity of azote and carbonic acid, as to prevent the explosion of the fire-damp, and which, by the nature of its apertures for giving admittance and exit to the air, should be rendered incapable of communicating any explosion to the external air.

" Having succeeded in the construction of safe lanterns and lamps, equally portable with common lanterns and lamps, which afforded sufficient light, and which bore motion perfectly well, I submitted them individually to practical tests, by throwing into them explosive atmospheres of fire-damp and air. By the natural action of the flame drawing air through the air canals, from the explosive atmosphere, the light was uniformly extinguished; and when an explosive mixture was forcibly pressed into the body of the lamp, the explosion was always stopped by the safety apertures, which may be said figuratively to act as a sort of *chemical fire sieves* in separating flame from air. But I was not contented with these trials, and I submitted the safe canals, tubes. and wire gauze fire sieves, to much more severe tests : I made them the medium of communication between a large glass vessel filled with the strongest explosive mixture of carburetted hydrogene and air, and a bladder two-thirds or one-half full of the same mixture, both insulated from the atmosphere. By means of wires passing near the stop-cock of the glass vessel, I fired the explosive mixture in it by the discharge of a Leyden jar. The bladder always expanded at the moment the explosion was made; a contraction as rapidly took place; and a lambent flame played round the mouths of the safety apertures, open in the glass vessel; but the mixture in the bladder did not explode: and by pressing some of it into the glass vessel, so as to make it replace the foul air, and subjecting it to the electric spark, repeated explosions were produced, proving the perfect security of the safety apertures; even when acted on by a much more powerful explosion than could possibly occur from the introduction of air from the mines.

These experiments held good whatever were the proportions of the explosive mixture and whatever was the size of the glass vessel, (no one was ever used containing more than a quart) provided as many as 12 metallic tubes were used of one-seventh of an inch in diameter, and two and a half inches long; or provided the circular metallic canals, were one-twenty-fifth of an inch in diameter, one and one-seventh of an inch deep, and at least two inches in circumference; or provided the wire gauze had apertures of only one-one-hundred-and-twentieth of an inch. When 12 metallic tubes were employed as the medium of communication, one-seventh of an inch in diameter and an inch long, the explosion was communicated by them into the bladder. Four glass tubes of the one-sixteenth of an inch in diameter and two inches long, did not communicate the explosion; but *one* of this diameter and length produced the effect. The explosion was stopped by a single tube one-twenty-eighth of an inch in diameter, when it was three inches long, but not when it was two inches long.

The explosion was stopped by the metallic gauze of one-one-hundred-and-twentieth when it was placed between the exploding vessel and the bladder, though it did not present a surface of more than half a square inch, and the explosive mixture in the bladder in passing through it to supply the vacuum produced in the glass vessel, burnt on the surface exposed to the glass vessel for some seconds, producing a murmuring noise.

A circular canal one-twenty-fifth of an inch in diameter, an inch and a half in circumference, and one and seven-tenths of an inch deep communicated explosion, but four concentric canals, of the same depth and diameter, and of which the smallest was two inches in circumference, and separated from each other only by their sides, which were of brass, and about one-fortieth of an inch in thickness, did not suffer the explosion to act through them.

It would appear then, that the smaller the circumference of the canal, that is the nearer it approaches to a tube, the greater must be its depth, or the less its diameter to render it safe.

I did not perceive any difference in these experiments, when the metals of the apertures were warmed by repeated explosions; it is probable, however, that considerable elevation of temperature would increase the power of the aperture to pass the explosion; but the difference between the temperature of flame, and that marked on our common mercurial scale, is so great that the addition of a few degrees of heat probably does not diminish perceptibly the cooling power of a metallic surface, with regard to flame.

By diminishing the diameter of the air canals their power of passing the explosion is so much diminished that their depth and circumference may be brought extremely low. I found that flame would not pass through a canal of the one-seventieth of an inch in diameter, when it was one-fourth of an inch deep, and forming a cylinder of only one-fourth of an inch in circumference; and a number of apertures of one-one-hundredth of an inch are safe when their depth is equal to their diameter. It is evident from these facts, that metallic doors, or joinings in lamps may be easily made safe by causing them to project upon and fit closely to parallel metallic surfaces.

Longitudinal air canals of metal may, I find, be employed with the same security as circular canals; and a few pieces of tin-plate soldered together with wires to regulate the diameter of the canal, answer the purpose of the feeder or safe chimney as well as drawn cylinders of brass.

A candle will burn in a lantern or glass tube made safe with metallic gauze, as well as in the open air; I conceive, however, that oil lamps, in which the wick will always stand at the same height, will be preferred.

But the principle applies to every kind of light, and its entire safety is demonstrated.

When the fire-damp is so mixed with the external atmosphere as to render it explosive, the light in the safe lantern or lamp will be extinguished, and warning will be given to the miners to withdraw from, and to ventilate that part of the mine.

It is probable, that when explosions occur from the sparks from the steel mill, the mixture of the fire-damp is in the proportion required to consume all the oxygene of the air, for it is only in about this proportion that explosive mixtures can be fired by electrical sparks from a common machine.

As the wick may be moved without communication between the air in the safe lantern or lamp and the atmosphere, there is no danger in trimming or feeding them; but they should be lighted in a part of the mine where there is no fire-damp, and by a person charged with the care of the lights; and by these inventions, used with such simple precautions, there is every reason to believe a number of lives will be saved, and much misery prevented. Where candles are employed in the open air in the mines, life is extinguished by the explosion; with the safe lantern or safe lamp the light is only put out, and no other inconvenience will occur.

It does not appear, by what I have learnt from the miners, that breathing an atmosphere containing a certain mixture of fire-damp near or even at the explosive point, is attended with any bad consequence. I ascertained that a bird lived in a mixture of equal parts of fire-damp and air; but he soon began to show symptoms of suffering. I found a slight head-ache produced by breathing for a few minutes an explosive mixture of fire-damp and air : and if merely the health of the miners be considered, the fire-damp ought always to be kept far below the point of its explosive mixture.

Miners sometimes are found alive in a mine after an explosion has taken place: this is easily explained, when it is considered that the inflammation is almost always limited to a particular spot, and that it mixes the residual air with much common air; and supposing 1 of fire-damp to 13 of air to be exploded, there will still remain nearly one-third of the original quantity of oxygene in the residual gas: and in some experiments, made 16 years ago, I found that an animal lived, though with suffering, for a short time, in a gas containing 100 parts of azote, 14 parts of carbonic acid, and 7 parts of oxygene."

2. *An Account of a Method for giving Light in* *explosive Mixtures of Fire-Damp in Coal Mines by* *consuming the Fire-Damp.** *Jan.* 1816.

" In this communication I shall describe a light that will burn in any explosive mixture of fire-damp, and the light of which arises from the combustion of the fire-damp itself.

" The invention consists in covering or surrounding a flame of a lamp or candle by a wire sieve ; the coarsest that I have tried with perfect safety contained 625 apertures in a square inch, and the wire was one-seventieth of an inch in thickness, the finest 6,400 apertures in a square inch, and the wire was one-two-hundred-and-fiftieth of an inch in diameter.

" When a lighted lamp or candle screwed into a ring soldered to a cylinder of wire gauze, having no apertures except those of the gauze, or safe apertures, is introduced into the most explosive mixture of carburetted hydrogen and air, the cylinder becomes filled with a bright flame, and this

* From Phil. Trans.

flame continues to burn as long as the mixture is explosive. When the carburetted hydrogen is to the air as 1 to 12, the flame of the wick appears within the flame of the fire damp, when the proportion is as high as 1 to 7, the flame of the wick disappears.

" When the thickest wires are used in the gauze it becomes strongly red hot, particularly at the top, but yet no explosion takes place. The flame is brighter the larger the apertures of the gauze, and the cylinder of 625 apertures to the square inch, gives a brilliant light in a mixture of 1 part of gas from the distillation of coal and 7 parts of air. The lower part of the flame is green, the middle purple, and the upper part blue.

" I have tried cylinders of 6400 apertures to the square inch, in mixtures of oxygen and carburetted hydrogen, and even in mixtures of oxygen and hydrogen, and though the wire became intensely red hot, yet explosions never took place; the combustion was entirely limited to the interior of the lamp.

" In all these experiments, there was a

noise like that produced by the burning of hydrogen gas in open tubes.

" These extraordinary and unexpected results lead to many enquiries, respecting the nature and communication of flame, but my object at present is only to point out their application to the use of the collier.

" All that he requires to ensure security, are small wire cages to surround his candle or his lamp, which may be made for a few pence, and of which various modifications may be adopted, and the application of this discovery will not only preserve him from the fire-damp, but enable him to apply it to use, and to destroy it at the same time that gives him a useful light."

3. *On the Combustion of explosive Mixtures confined by Wire Gauze.**

" I HAVE pursued my enquiries respecting the limits of the size of the apertures and of the wire in the metallic gauze, which I have applied to secure the coal miners from the

* From PHIL. TRANS.

explosions of fire-damp. Gauze made of brass wire, one-fiftieth of an inch in thickness, and containing only ten apertures to the inch, or 100 apertures in the square inch, employed in the usual way as a guard of flame, did not communicate explosion in a mixture of 1 part of coal gas and 12 parts of air, as long as it was cool, but as soon as the top became hot, an explosion took place.

" A quick lateral motion likewise enabled it to communicate explosion.

" Gauze made of the same wire, containing 14 apertures to the inch, or 196 to the square inch, did not communicate explosion till it became strongly red hot, when it was no longer safe in explosive mixtures of coal gas; but no motion that could be given to it, by shaking it in a close jar, produced explosion.

" Iron wire gauze of one-fortieth, and containing 240 apertures in the square inch, was safe in explosive mixtures of coal gas, till it became strongly red hot at the top.

" Iron wire gauze of one-fiftieth, and of 24 apertures to the inch, or of 576 to the square inch, appeared safe under all circumstances in explosive mixtures of coal gas. I kept up a continual flame in a cylinder of this kind, 8 inches high and 2 inches in diameter, for a quarter of an hour, varying the proportions of coal gas and air as far as was compatible with their inflammation ; the top of the cylinder, for some minutes, was strongly red hot, but though the mixed gas was passed rapidly through it by pressure from a gasometer and a pair of double bellows, so as to make it a species of blast furnace, yet no explosion took place.

" I mentioned in my last communication to the Society, that a flame confined in a cylinder of very fine wire gauze, did not explode a mixture of oxygene and hydrogene, but that the gases burnt in it with great vivacity. I have repeated this experiment in nearly a pint of the most explosive mixture of the two gases; they burnt violently within the cylinder, but, though the upper part became nearly white hot, yet no explosion was communicated, and it was necessary to withdraw the cylinder to prevent the brass wire from being melted.

" These results are best explained by

considering the nature of the flame of combustible bodies, which, in all cases, must be considered as the combustion of an *explosive mixture* of inflammable gas, or vapour and air; for it cannot be regarded as a mere combustion at the surface of contact of the inflammable matter: and the fact is proved by holding a taper or a piece of burning phosphorus within a large flame made by the combustion of alcohol, the flame of the candle or of the phosphorus will appear in the centre of the other flame, proving that there is oxygene even in its interior part.

" The heat communicated by flame must depend upon its mass; this is shown by the fact, that the top of a slender cylinder of wire gauze hardly ever becomes dull red in the experiment on an explosive mixture, whilst in a larger cylinder, made of the same material, the central part of the top soon becomes bright red. A large quantity of cold air thrown upon a small flame, lowers its heat beyond the explosive point, and in extinguishing a flame by blowing upon it, the effect is probably principally produced by this cause.

" If a piece of wire gauze sieve is held over a flame of a lamp or of coal gas, it prevents the flame from passing it, and the phenomenon is precisely similar to that exhibited by the wire gauze cylinders; the air passing through is found very hot, for it will convert paper into charcoal; and it is an explosive mixture, for it will inflame if a lighted taper be presented to it, but it is cooled below the explosive point by passing through wires even red hot, and by being mixed with a considerable quantity of air comparatively cold. The real temperature of visible flame is perhaps as high as any we are acquainted with. Mr. TENNANT was in the habit of showing an experiment, which demonstrates the intensity of its heat. He used to fuse a small filament of platinum in the flame of a common candle; and it is proved by many facts, that a stream of air may be made to render a metallic body white hot, yet not be itself luminous.

" A considerable mass of heated metal is required to inflame even coal gas, or the contact of the same mixture with an extensive heated surface. An iron wire of

one-twentieth of an inch and 8 inches long
red hot, when held perpendicularly in a
stream of coal gas, did not inflame it, nor
did a short wire of one-sixth of an inch
produce the effect held horizontally; but
wire of the same size, when six inches of it
were red hot, and when it was held perpendicularly in a bottle, containing an explosive mixture, so that heat was successively
communicated to portions of the gas, produced its explosion."

4. *Some new Views and Experiments respecting Flame.* *

WHEN a wire gauze safe lamp is made
to burn in a very explosive mixture of coal
gas and air, the light is feeble, and of a
pale colour; whereas the flame of a current
of coal gas burnt in the atmosphere, as is
well known by the phenomena of the gas
lights, is extremely brilliant. I have en-

* From the Journal of Science and the Arts.

deavoured to shew, page 46, that in all cases
flame is a continued combustion of explosive
mixtures ; it becomes, therefore, a problem
of some interest, " Why the combustion of
explosive mixtures, under different circumstances, should produce such different appearances ?" A very acute philosopher, who
himself started the subject in conversation,
suggested the idea, that in the combustion
of explosive mixtures within the lamp, carbonic oxide might be formed ; and that the
light might be deficient, from the deficiency
of the quantity of oxygene necessary to
produce carbonic acid. On submitting this
idea to the test of experiment, it was discovered to be unfounded ; for, by the combustion in the wire gauze lamp, carbonic
acid was produced in quantities as great
as could have been expected from the
quantity of oxygene consumed ; and on
adding oxygene to a mixture in quantities
more than sufficient to burn the whole of
the gas, the character of the light still continued the same.

In reflecting on the circumstances of the
two species of combustion, I was led to
imagine that the cause of the superiority of

the light of the *stream* of coal gas might be owing to the *decomposition* of a part of the gas towards the interior of the flame where the air was in smallest quantity, and the deposition of solid charcoal, which, first by its *ignition*, and afterwards by its *combustion*, increased in a high degree the intensity of the light; and a few experiments soon convinced me that this was the true solution of the problem.

I held a piece of wire gauze, of about 900 apertures to the square inch, over a stream of coal gas issuing from a small pipe, and inflamed the gas above the wire gauze, which was almost in contact with the orifice of the pipe; when it burned with its usual bright light. On raising the wire gauze so as to cause the gas to be mixed with more air before it inflamed, the light became feebler; and at a certain distance the flame assumed the precise character of that of an explosive mixture burning within the lamp; but though the light was so feeble in this last case, the heat was greater than when the light was much more vivid, and a piece of wire of platinum held in this feeble blue flame became instantly white hot.

On reversing the experiment by inflaming a stream of coal gas, and passing a piece of wire gauze gradually from the summit of the flame to the orifice of the pipe, the result was still more instructive; for it was found that the apex of the flame intercepted by the wire gauze afforded no solid charcoal; but in passing it downwards, solid charcoal was given off in considerable quantities, and prevented from burning by the cooling agency of the wire gauze; and at the bottom of the flame, where the gas burnt blue in its immediate contact with the atmosphere, charcoal ceased to be deposited in visible quantities.

This principle of the increase of the brilliancy and density of flame by the production and ignition of solid matter, appears to admit of many applications.

1st. It explains readily the appearances of the different parts of the flames of burning bodies, and of flame urged by the blow-pipe : the point of the inner blue flame, where the heat is greatest, is the point where the whole of the charcoal is burnt in its gaseous combinations without previous deposition.

2dly. It explains the intensity of the light

of those *flames* in which *fixed* solid matter is produced in combustion, such as that of the flame of phosphorus* and of zinc in oxygene, &c. and of potassium in chlorine; and the feebleness of the light of those flames in which gaseous and volatile matter alone is produced, such as those of hydrogene and sulphur in oxygene, phosphorus in chlorine, &c.

3rdly. It offers means of increasing the light of certain burning substances, by placing in their flames even incombustible substances. Thus the intensity of the light of burning sulphur, carbonic oxide, &c. is wonderfully increased by throwing into them oxide of zinc, or by placing in them very fine amianthus or metallic gauze.

4thly. It leads to deductions respecting the chemical nature of bodies and various phenomena of their decomposition. Thus ether burns with a flame which seems to indicate the presence of olefiant gas in that substance. Alcohol burns with a flame si-

---

* Since this paper has been written I have found that phosphoric acid volatilizes slowly at a strong red heat, but under moderate pressure it bears a white heat, and in a flame so intense as that of phosphorus, the elastic force must produce the effect of compression.

milar to that of a mixture of carbonic oxide
and hydrogene ; so that the first is probably
a binary compound of olefiant gas and
water, and the second of carbonic oxide
and hydrogene.

When cuprane or protochlorid of copper
is introduced into the flame of a candle or
lamp, it affords a peculiar dense and brilliant red light, tinged with green and blue
towards the edges, which seems to depend
upon the chlorine being separated from the
copper by the hydrogene, and the ignition
and combustion of the solid copper and
charcoal.

Similar explanations may be given of
the phenomena presented by the action of
other combinations of chlorine on flame ;
and it is probable, in many of those cases
when the colour of flame is changed by
the introduction of incombustible compounds, that the effect depends upon the
production and subsequent ignition or combustion of inflammable matter from them.
Thus the rose-coloured light given to flame
by the compounds of strontium and calcium,[*]

[*] A similar effect I find is produced by the compounds
of the new fixed alkali lithia which serves to distinguish
its compounds from those of Potassa and Soda.—1818.

and the yellow colour given by those of barium, and the green by those of boron, may depend upon a temporary production of these bases by the inflammable matter of the flame.

Whenever a flame is remarkably brilliant and dense, it may be always concluded that some solid matter is produced in it: on the contrary, when a flame is extremely feeble and transparent, it may be inferred that no solid matter is formed. Thus none of the volatile combinations of sulphur burn with a flame in the slightest degree opaque ; and, consequently, there is no reason, from the phenomena of its flame, to suspect the existence of any fixed basis in sulphur.

5thly. These views will probably offer illustrations of electrical light. The voltaic arc of flame from the great battery, differs in colour and intensity according as the substances employed in the circuit are different; and is infinitely more brilliant and dense with charcoal than with any other substance. May not this depend upon particles of the substances separated by the electrical attractions ? and the particles of charcoal being the lightest amongst solid bodies, (as their elementary proportional number shews,) and the least coherent, would be separated in the largest quantities.

6thly. The heat of flames may be actually diminished by increasing their light, (at least the heat communicable to other matter,) and vice versa. The flame from combustion which produces the most intense heat amongst those I have examined, is that of a mixture of oxygene and hydrogene in slight excess, compressed in a blow-pipe apparatus, and inflamed from a tube having a very small aperture.* This flame is hardly visible in bright day-light, yet it instantly fuses very refractory bodies; and the light from solid matters ignited in it, is so vivid as to be painful to the eye.

July, 1816.

" * John George Children, Esq. first proposed to me this application of the blow-pipe by compression, immediately after I had discovered that the explosion from oxygene and hydrogene might be arrested by very small apertures, and I first tried the experiment with a fine glass capillary tube. The flame was *not visible* at the end of this tube, being overpowered by the brilliant star of the glass ignited at the aperture." The results produced by the action of this flame on solid bodies, were so similar to those published by Mr. Hare some years ago, that I did not think them worthy of particular notice.

5. *Some additional Researches on Flame.**
Jan. 1817.

" THAT greater distinctness may exist in
these details, I shall treat of my subjects
under four heads. In the first I shall dis-
cuss the effects of rarefaction, by partly
removing the pressure of the atmosphere
upon flame and explosion. In the second,
I shall consider the effects of heat in combustion. In the third, I shall examine the
effect of the mixture of gaseous substances
not concerned in combustion upon flame
and explosion. In the fourth, I shall offer
some general views upon flame, and point
out certain practical and theoretical applications of the results.

1. On the effect of rarefaction by partly removing the
pressure of the atmosphere upon flame and explosion.

THE earlier experimenters upon the
BOYLEAN vacuum observed that flame
ceased in highly rarefied air : but the degree
of rarefaction necessary for this effect, has
been differently stated. Amongst late ex-

* PHIL. TRANS.

perimenters, M. de GROTTHUS has examined this subject. He has asserted that
a mixture of oxygene and hydrogene ceases
to be explosive by the electrical spark when
rarefied sixteen times, and that a mixture
of chlorine and hydrogene cannot be exploded when rarefied only six times, and he
generalises by supposing that rarefaction,
whether produced by removing pressure or
by heat, has the same effect.

I shall not begin by discussing the experiments of this ingenious author. My
own results and conclusions are very different from his; and the cause of this difference, will I think be obvious in the course
of these inquiries. I shall proceed in stating
the observations which guided my researches.

When hydrogene gas slowly produced
from a proper mixture was inflamed at a
fine orifice of a glass tube, as in the experiment called the philosophical candle, so as
to make a jet of flame of about one-sixth of
an inch in height, and introduced under the
receiver of an air pump containing from
200 to 300 cubical inches of air, the flame
enlarged as the receiver became exhausted;

and when the gage indicated a pressure between 4 and 5 times less than that of the atmosphere was at its maximum of size, it then gradually diminished below, but burned above till the pressure was between 7 and 8 times less, when it became extinguished.

To ascertain whether the effect depended upon the deficiency of oxygene, I used a larger jet with the same apparatus, when the flame to my surprise burned longer, and when the atmosphere was rarefied ten times, and this in repeated trials. When the larger jet was used, the point of the glass tube became white hot, and continued red hot till the flame was extinguished. It immediately occurred to me, that the heat communicated to the gas by this tube, was the cause that the combustion continued longer, in the last trials when the larger flame was used; and the following experiments confirmed the conclusion. A piece of wire of platinum was coiled round the top of the tube, so as to reach into and above the flame. The jet of gas of one-sixth of an inch in height was lighted and the exhaustion made; the wire of platinum soon became white hot in the centre of the flame, and a small point of wire near the top fused : it continued white hot till the pressure was 6 times less, when it was 10 times it continued red hot at the upper part, and as long as it was dull red, the gas though extinguished below, continued to burn in contact with the hot wire, and the combustion did not cease until the pressure was reduced 13 times.

It appears from this result, that the flame of hydrogene is extinguished in rarefied atmospheres, only when the heat it produces is insufficient to. keep up the combustion, which appears to be when it is incapable of communicating visible ignition to metal, and as this is the temperature required for the inflammation of hydrogene at common pressures, it appears that its *combustibility* is neither diminished nor increased by rarefaction from the removal of pressure.

According to this view with respect to hydrogene, it should follow that amongst other combustible bodies, those which require least heat for their combustion, ought to burn in more rarefied air than those that require more heat, and those that produce much heat in their combustion ought to burn, other circumstances being the same, in more rarefied air than those that produce little heat : and every experiment I have made confirms these conclusions. Thus olefiant gas which approaches nearly to hydrogene in the heat produced by its combustion, and which does not require a much higher temperature for its inflammation, when its flame was made by a jet of gas from a bladder connected with a small tube furnished with a wire of platinum, under the same circumstances as hydrogene, ceased to burn when the pressure was diminished between 10 and 11 times : and the flames of alcohol and of the wax taper which require a greater consumption of heat for the volatilization and decomposition of their combustible matter, were extinguished when the pressure was five or six times less without the wire of platinum, and 7 or 8 times less when the wire was kept in the flame. Light carburetted hydrogene, which produces, as will be seen hereafter, less heat in combustion than any of the common combustible gases, except carbonic oxide, and which requires a higher temperature for its inflammation than any other, had its flame extinguished, even though the tube was furnished with the wire when the pressure was below one-fourth.

The flame of carbonic oxide which, though it produces little heat in combustion, is as inflammable as hydrogene, burned when the wire was used, the pressure being one-sixth.

The flame of sulphuretted hydrogene, the heat of which is in some measure carried off by the sulphur produced by its decomposition during its combustion in rare air, when burned in the same apparatus as the olefiant and other gases, was extinguished when the pressure was one-seventh.

Sulphur, which requires a lower temperature for its combustion than any common inflammable substance, except phosphorus, burned with a very feeble blue flame in air rarefied fifteen times, and at this pressure the flame heated a wire of platinum to dull redness, nor was it extinguished till the pressure was reduced to one-twentieth.*

* The temperature of the atmosphere diminishes in a certain ratio with its height, which must be attended to in the conclusions respecting combustion in the upper regions of the atmosphere, and the elevation must be

Phosphorus, as has been shown by M. VAN MARUM, burns in an atmosphere rarefied 60 times; and I found that phosphuretted hydrogene produced a flash of light when admitted into the best vacuum that could be made, by an excellent pump of NAIRN'S construction.

The mixture of chlorine and hydrogene inflames at a much lower temperature than that of hydrogene and oxygene, and produces a considerable degree of heat in combustion; it was therefore probable that it would bear a greater degree of rarefaction, without having its power of exploding destroyed; and this I found by many trials is actually the case, contrary to the assertion of M. de GROTTHUS. Oxygene and hydrogene in the proportion to form water, will not explode by the electrical spark when rarefied eighteen times, but hydrogene and chlorine in the proportion to form mu-

somewhat lower than in arithmetical progression, the pressure decreasing in geometrical progression.

There is, however, every reason to believe, that the taper would be extinguished at a height of between 9 and 10 miles, hydrogene between 12 and 13, and sulphur between 15 and 16.

riatic acid gas, gave a distinct flash of light under the same circumstances, and they combined with visible inflammation when the spark was passed through them, the exhaustion being to one-twenty-fourth.

The experiment on the flame of hydrogene with the wire of platinum, and which holds good with the flames of the other gases, shows, that by preserving heat in rarefied air, or giving heat to a mixture, inflammation may be continued when, under common circumstances, it would be extinguished. This I found was the case in other instances, when the heat was differently communicated: thus, when camphor was burned in a glass tube, so as to make the upper part of the tube red hot, the inflammation continued when the rarefaction was 9 times, whereas it would only continue in air rarefied 6 times, when it was burned in a thick metallic tube which could not be considerably heated by it.

By bringing a little naphtha in contact with red hot iron, it produced a faint lambent flame, when there remained in the receiver only one-thirtieth of the original quantity of air, though without foreign heat its flame was extinguished when the quantity was one-sixth.

I rarefied a mixture of oxygene and hydrogene by the air pump to about eighteen times, when it could not be inflamed by the electric spark. I then heated strongly the upper part of the tube till the glass began to soften, and passed the spark, when a feeble flash was observed not reaching far into the tube, the heated gases only appearing to enter into inflammation. This last experiment requires considerable care. If the exhaustion be much greater, or if the heat be raised very slowly,* it does not succeed; and if the heat be raised so high as to make the glass luminous, the flash of light, which is extremely feeble, is not visible: it is difficult to procure the proper degree of exhaustion, and to give the exact degree of heat; I have, however, succeeded three times in obtaining the results, and in one instance it was witnessed by Mr. BRANDE.

To elucidate the enquiry still farther,

* The reason will be obvious from what will be stated page 72.

I made a series of experiments on the heat produced by some of the inflammable gases in combustion. In comparing the heat communicated to wires of platinum by flames of the same size, it was evident, that hydrogene and olefiant gas in oxygene, and hydrogene in-chlorine, produced a much greater intensity of heat in combustion, than the other gaseous substances I have named burned in oxygene : but no regular scale could be formed from observations of this kind. I endeavoured to gain some approximations on the subject by burning equal quantities of different gases under the same circumstances, and applying the heat to an apparatus by which it could be measured. For this purpose a mercurial gas holder was furnished with a system of stop cocks, terminating in a strong tube of platinum having a minute aperture. Above this was fixed a copper cup filled with olive oil, in which a thermometer was placed. The oil was heated to $212°$ to prevent any differences in the communication of heat by the condensation of aqueous vapour; the pressure was the same for the different gases and they were consumed as nearly as possible in the same time, and the flame ap-

F

plied to the same point of the copper cup, the bottom of which was wiped after each experiment.

The results were as follows:

The flame from olefiant gas raised the

thermometer to - 270°

―――――――― hydrogene - - 238

―――――――― sulphuretted hydrogene 232

―――――――― coal gas - - 236

―――――――― gaseous oxide of carbon 218

The quantities of oxygene consumed (that absorbed by the hydrogene being taken as 1) would be, supposing the combustion perfect, for the olefiant gas 6, for the sulphuretted hydrogene 3, for the carbonic oxide 1. The coal gas contained only a very small proportion of olefiant gas; supposing it to be pure carburetted hydrogene, it would have consumed 4 of oxygene. Taking the elevations of temperature, and the quantities of oxygene consumed as the data, the ratios of the heat produced by the combustion of the different gases, would be for hydrogene 26, for olefiant gas 9.66, for sulphuretted hydrogene 6.66, for carburetted hydrogene 6, for carbonic oxide 6*.

* These results may be compared with Mr. DALTON's

It will be useless to reason upon this ratio as exact, for charcoal was deposited both from the olefiant gas and coal gas during the experiment, and much sulphur was deposited from the sulphuretted hydrogene ; and there is great reason to believe, that the capacities of fluids for heat increase with their temperature. It confirms, however, the general conclusions, and proves that hydrogene stands at the head of the scale, and gaseous oxide of carbon at the bottom. It might at first view be imagined that, according to this scale, the flame of carbonic oxide ought to be extinguished by rarefaction, at the same degree as that of carburetted hydrogene ; but it must be remembered, as I have mentioned in another place, that carbonic oxide is a much more combustible gas. Carbonic oxide inflames in the atmosphere when brought into contact with an iron wire heated to dull redness, whereas carburetted hydrogene is not inflammable by a similar wire, unless it is heated to whiteness so as to burn with sparks.

new system of Chemical Philosophy ; they agree in shewing that hydrogene produces more heat in combustion than any of its compounds.

2. On the effects of rarefaction by heat on combustion and explosion.

The results detailed in the preceding section are indirectly opposed to the opinion of M. DE GROTTHUS, that rarefaction by heat destroys the combustibility of gaseous mixtures. Before I made any direct experiments on this subject, I endeavoured to ascertain the degree of expansion which can be communicated to elastic fluids by the strongest heat that can be applied to glass vessels. For this purpose I introduced into a graduated curved glass tube some fusible metal. I heated the fusible metal and the portion of the tube containing the air included by it, under boiling water for some time. I then placed the apparatus in a charcoal fire, and very gradually raised the temperature till the fusible metal appeared luminous when viewed in the shade. At this time the air had expanded so as to occupy 2.25 parts in the tube, it being 1 at the temperature of boiling water. Another experiment was made in a thicker glass tube, and the heat was raised until the tube began to run together ; but though this heat appeared cherry red, the expansion was not to more than 2.5, and a part of this might perhaps have been apparent only, owing to the collapsing of the glass tube before it actually melted. It may be supposed that the oxidation of the fusible metal may have had some effect in making the expansion appear less ; but in the first experiment the air was gradually brought back to its original temperature of boiling water, when the absorption was scarcely sensible. If M. GAY LUSSAC's conclusions be taken as the ground work of calculation, and it be supposed that air expands equally for equal increments of temperature, it would appear that the temperature of air capable of rendering glass luminous must be 1035° Fahrenheit.*

M. DE GROTTHUS describes an experiment in which atmospheric air and hydrogene, expanded to four times their bulk over

* The mode of ascertaining temperatures as high as the point of fusion of glass by the expansion of air, seems more unexceptionable than any other. It gives for the point of visible ignition nearly the same degree as that deduced by NEWTON from the times of the cooling of ignited metal in the atmosphere.

mercury by heat, would not inflame by the electric spark. It is evident, that in this experiment a large quantity of steam or of mercurial vapour must have been present, which like other inexplosive elastic fluids, prevents combustion when mixed in certain quantities with explosive mixtures; but though he seems aware that his gases were not dry, yet he draws his general conclusion, that expansion by heat destroys the explosive powers of gases, principally from this inconclusive experiment.

I introduced into a small graduated tube over well boiled mercury, a mixture of two parts of hydrogene and one of oxygene, and heated the tube by a large spirit lamp till the volume of the gas was increased from 1 to 2.5. I then, by means of a blow pipe and another spirit lamp, made the upper part of the tube red hot, when an explosion instantly took place.

I introduced into a bladder a mixture of oxygene and hydrogene, and connected this bladder with a thick glass tube of about one-sixth of an inch in diameter and three feet long, curved so that it could be gradually heated in a charcoal furnace; two spirit lamps were placed under the tube where it entered the charcoal fire, and the mixture was very slowly pressed through : an explosion took place before the tube was red hot.

This experiment shows that expansion by heat, instead of diminishing the combustibility of gases on the contrary, enables them to explode apparently at a lower temperature, which seems perfectly reasonable, as a part of the heat communicated by any ignited body must be lost in gradually raising the temperature. I made several other experiments which establish the same conclusions. A mixture of common air and hydrogene was introduced into a small copper tube, having a stopper not quite tight; the copper tube was placed in a charcoal fire; before it became visibly red an explosion took place, and the stopper was driven out.

I made various experiments on explosions by passing mixtures of hydrogene and oxygene through heated tubes; in the beginning of one of these trials, in which the heat was much below redness, steam appeared to be formed without any combustion. This led me to expose mixtures of oxygene and hydrogene in tubes, in which they were confined by fluid fusible metal to heat; and I found that by carefully applying a heat between the boiling point of mercury, which is not sufficient for the effect, and a heat approaching to the greatest heat that can be given without making glass luminous in darkness, the combination was effected without any violence and without any light: and commencing with $212°$, the volume of steam formed at the point of combination. appeared exactly equal to that of the original gases. So that the first effect in experiments of this kind is an expansion, afterwards a contraction, and then the restoration of the primitive volume.

If when this change is going on, the heat be quickly raised to redness, an explosion takes place; but with small quantities of gas the change is completed in less than a minute.

It is probable, that the slow combination without combustion, already long ago observed with respect to hydrogene and chlorine, oxygene and metals, will happen at certain temperatures with most substances that unite by heat. On trying charcoal, I found that at a temperature which appeared to be a little above the boiling point of quicksilver, it converted oxygene pretty rapidly into carbonic acid, without any luminous appearance, and at a dull red heat, the elements of olefiant gas combined in a similar manner with oxygene, slowly and without explosion.

The effect of the slow combination of oxygene and hydrogene is not connected with their rarefaction by heat, for I found that it took place when the gases were confined in a tube by fusible metal rendered solid at its upper surface; and certainly as rapidly, and without any appearance of light.

M. DE GROTTHUS has stated, that, if a glowing coal be brought into contact with a mixture of oxygene and hydrogene, it only rarefies them, but does not explode them: but this depends upon the degree of heat communicated by the coal: if it is red in day light and free from ashes, it uniformly explodes the mixture; if its redness is barely visible in shade, it will not explode them, but cause their slow combination: and the general phenomenon is wholly unconnected with rarefaction, as is shown by the following circumstance. When the heat is greatest, and before the invisible combination is completed, if an iron wire heated to to whiteness be placed upon the coal within the vessel, the mixture instantly explodes.

Light carburetted hydrogene, or pure fire-damp, as has been shown, requires a very strong heat for its inflammation; it therefore offered a good substance for an experiment on the effect of high degrees of rarefaction by heat on combustion. I mixed together one part of this gas and eight parts of air, and introduced them into a bládder furnished with a capillary tube. I heated this tube till it began to melt, and then slowly passed the mixture through it into the flame of a spirit lamp, when it took fire and burned with its own peculiar explosive light beyond the flame of the lamp, and when withdrawn, though the aperture was quite white hot, it continued to burn vividly.

That the compression in one part of an explosive mixture produced by the sudden expansion of another part by heat, or the electric spark, is not the cause of combi-

nation, as has been supposed by Dr. Hig-
gins, M. Berthollet, and others, appears
to be evident from what has been stated,
and it is rendered still more so by the following facts. A mixture of hydro-phosphoric
gas (bi-phosphuretted hydrogene gas) and
oxygene, which explode at a heat a little
above that of boiling water, was confined by
mercury, and very gradually heated on a
sand bath; when the temperature of the
mercury was 242°, the mixture exploded.

A similar mixture was placed in a receiver communicating with an indensing
syringe, and condensed over mercury till it
occupied only one-fifth of its original
volume. No explosion took place, and no
chemical change had occurred, for when its
volume was restored, it was instantly exploded by the spirit lamp.

It would appear, then, that *the heat* given
out by the compression of gases is the real
cause of the combustion which it produces,
and that at certain elevations of temperature,
whether in rarefied or compressed atmospheres, explosion or combustion occurs, i. e.
bodies combine with the production of heat
and light.

3. On the effects of the mixture of different gases in explosion and combustion.

In my first Paper on the fire-damp of
coal mines, I have mentioned that carbonic
acid gas has a greater power of destroying
the explosive power of mixtures of fire-damp
and air than azote, and I have ventured to
suppose the cause to be its greater density
and capacity for heat, in consequence of
which it might exert a greater cooling
agency, and prevent the temperature of the
mixture from being raised to that degree
necessary for combustion. I have lately
made a series of experiments with the view
of determining how far this idea is correct,
and for the purpose of ascertaining the
general phenomena of the effects of the
mixture of gaseous substances upon explosion and combustion.

I took given volumes of a mixture of
two parts of hydrogene and one part of oxygene by measure, and diluting them with
various quantities of different elastic fluids,
I ascertained at what degree of dilution the
power of inflammation by a strong spark
from a Leyden phial was destroyed. I

found that for one of the mixture inflammation was prevented by

Of Hydrogene, about - -	8
Oxygene - - -	9
Nitrous oxide - -	11
Carburetted hydrogene -	1
Sulphuretted Hydrogene -	2
Olefiant gas - - -	$\frac{1}{2}$
Muriatic acid gas - -	2
Silicated fluoric acid gas -	$\frac{1}{6}$

Inflammation took place when the mixtures contained of

Hydrogene - - -	6
Oxygene - - - -	7
Nitrous oxide - - -	10
Carburetted hydrogene -	$\frac{3}{4}$
Olefiant gas - - -	$\frac{1}{3}$
Sulphuretted hydrogene -	$1\frac{1}{2}$
Muriatic acid gas - -	$1\frac{1}{2}$
Fluoric acid gas - - -	$\frac{3}{4}$

I hope to be able to repeat these experiments with more precision at no distant time; the results are not sufficiently exact to lay the foundation for any calculations on the relative cooling powers of equal volumes of the gases, but they show sufficiently, if the conclusions of M. M. DE LA ROCHE and BERARD be correct, that other causes, besides density and capacity for heat, interfere with the phenomena. Thus nitrous oxide, which is nearly one-third denser than oxygene, and which, according to DE LA ROCHE and BERARD, has a greater capacity for heat in the ratio of 1.3503 to 9765 in volume, has lower powers of preventing explosion ; and hydrogene, which is 15 times lighter than oxygene, and which in equal volumes has a smaller capacity of heat, certainly has a higher power of preventing explosion; and olefiant gas exceeds all other gaseous substances in a much higher ratio than could have been expected from its density and capacity. The olefiant gas I used was recently made, and might have contained some vapour of ether, and the nitrous oxide was mixed with some azote, but these slight causes could not have interfered with the results to any considerable extent.

Mr. LESLIE, in his elaborate and ingenious researches on heat, has observed the high powers of hydrogene of abstracting heat from solid bodies, as compared with that of common air and oxygene. I made a few experiments on the comparison of the powers of hydrogene, in this respect, with those of carburetted hydrogene, azote, oxygene, olefiant gas, nitrous oxide, chlorine, and carbonic acid gas. The same thermometer raised to the same temperature, 160°, was exposed to equal volumes (21 cubic inches) of olefiant gas, coal gas, carbonic acid gas, chlorine, nitrous oxide gas, hydrogene, oxygene, azote, and air, at equal temperatures. 52° Fahrenheit.

The times required for cooling to 106° were for

				′	″
Air	-	-	-	2	
Hydrogene		-	-		45
Olefiant gas		-	-	1.15	
Coal gas		-	-		55
Azote	-	●	-	1.30	
Oxygene		-	-	1.47	
*Nitrous oxide			-	2.30 2.53	
*Carbonic acid gas		-		2.45	
Chlorine		-	-	3.6	

It appears from these experiments, that

* These two last results were observed by Mr. FARADAY of the Royal Institution, when I was absent from the Laboratory.

the powers of elastic fluids to abstract or conduct away heat from solid surfaces, is in some inverse ratio to their density, and that there is something in the constitution of the light gases, which enables them to carry off heat from solid surfaces in a different manner from that in which they would abstract it in gaseous mixtures, depending probably upon the mobility of their parts.* The heating of gaseous media by the contact of fluid or solid bodies, as has been shown by Count RUMFORD, depends principally upon the change of place of their particles; and it is evident from the results stated in the beginning of this section, that these particles have different powers of abstracting heat analogous to the different powers of solids and fluids. Where an elastic fluid exerts a cooling influence on a solid surface, the effect must depend principally upon the rapidity with which its particles change their places: but where the

---

* Those particles which are lightest must be conceived most capable of changing place, and would therefore cool solid surfaces most rapidly : in the cooling of gaseous mixtures, the mobility of the particles can be of little consequence.

cooling particles are mixed throughout a mass with other gaseous particles, their effect must principally depend upon the power they possess of rapidly abstracting heat from the contiguous particles; and this will depend probably upon two causes, the simple abstracting power by which they become quickly heated, and their capacity for heat which is great in proportion as their temperatures are less raised by this abstraction.

Whatever be the cause of the different cooling powers of the different elastic fluids in preventing inflammation, very simple experiments show that they operate uniformly with respect to the different species of combustion, and that those explosive mixtures, or inflammable bodies, which require least heat for their combustion, require larger quantities of the different gases to prevent the effect, and *vicè versa;* thus one of chlorine and one of hydrogene still inflame when mixed with eighteen times their bulk of oxygene, whereas a mixture of carburetted hydrogene and oxygene in the proper proportions for combinations, one and two, have their inflammation prevented by less than three times their volume of oxygene.

A wax taper was instantly extinguished in air mixed with one-tenth of silicated fluoric acid gas, and in air mixed with one-sixth of muriatic acid gas; but the flame of hydrogene burned readily in those mixtures, and in mixtures in which the flame of hydrogene was extinguished, the flame of sulphur burned.

There is a very simple experiment which demonstrates in an elegant manner this general principle. Into a long bottle with a narrow neck introduce a lighted taper, and let it burn till it is extinguished; carefully stop the bottle, and introduce another lighted taper, it will be extinguished before it reaches the bottom of the neck: then introduce a small tube containing zinc and diluted sulphuric acid, and at the aperture of which the hydrogene is inflamed; the hydrogene will be found to burn in whatever part of the bottle the tube is placed: after the hydrogene is extinguished, introduce lighted sulphur; this will burn for some time, and after its extinction, phosphorus will be as luminous as in the air, and, if heated in the bottle, will produce a pale yellow flame of considerable density.

In cases when the heat required for chemical union is very small, as in the instance of hydrogene and chlorine, a mixture which prevents inflammation will not prevent combination, i. e. the gases will combine without any flash. This I witnessed in mixing two volumes of carburetted hydrogene with one of chlorine and hydrogene; muriatic acid was formed throughout the mixture, and heat produced, as was evident from the expansion when the spark passed, and the rapid contraction afterwards, but the heat was so quickly carried off by the quantity of carburetted hydrogene that no flash was visible.

In the case of phosphorus, which is combustible at the lowest temperature of the atmosphere, no known admixture of elastic fluid prevents the luminous appearance; but this seems to depend upon the light being limited to the solid particles of phosphoric acid formed; whereas to produce flame, a certain mass of elastic fluid must be luminous; and there is every reason to believe, that when phosphuretted hydrogene explodes in very rare air, it is only the phosphorus which is consumed. Any other substance that produces solid matter in combustion would probably be luminous in air as rare, or in mixtures as diluted, as phosphorus, provided the heat was elevated sufficiently for its combustion. I have found that this is actually the case with respect to zinc. I threw some zinc filings into an ignited iron crucible fixed on the stand of an air pump under a receiver, and exhausted until only one-sixtieth of the original quantity of air remained. When I judged that the red hot crucible must be full of the vapour of zinc, I admitted about one-sixtieth more of air, when a bright flash of light took place in and above the crucible, similar to that which is produced by admitting air to the vapour of phosphorus in vacuo.

The cooling power of mixtures of elastic fluids in preventing combustion must increase with their condensation, and diminish with their rarefaction; at the same time, the quantity of matter entering into combustion in given spaces, is relatively in‑creased and diminished. The experiments

on flame in rarefied atmospherical air, show
that the quantity of heat produced in combustion is very slowly diminished by rarefaction, the diminution of the cooling power
of the azote being apparently in a higher
ratio than the diminution of the heating
powers of the burning bodies. I endeavoured
to ascertain what would be the effect of
condensation on flame in atmospheric air,
and whether the cooling power of the azote
would increase in a lower ratio, as might be
expected, than the heat produced by the
increase of the quantity of matter entering
into combustion, but I found considerable
difficulties in making the experiments with
precision. I ascertained, however, that
both the light and heat of the flames of the
taper, of sulphur and hydrogene, were increased by acting on them by air condensed
four times; but not more than they would
have been by an addition of one-fifth of
oxygene.

I condensed air nearly five times, and
ignited iron wire to whiteness in it by the
voltaic apparatus, but the combustion took
place with very little more brightness than
in the common atmosphere, and would not

continue as in oxygene, nor did charcoal burn much more brightly in this compressed air than in common air. I intend to repeat these experiments, if possible, with higher condensing powers; they show sufficiently that, (for certain limits at least) as rarefaction does not diminish considerably the heat of flame in atmospherical air, so neither does condensation considerably increase it; a circumstance of great importance in the constitution of our atmosphere, which at all the heights or depths at which man can exist, still preserves the same relations to combustion.

It may be concluded from the general law, that at high temperatures, gases not concerned in combustion will have less powers of preventing that operation, and likewise, that steam and vapours, which require a considerable heat for their formation, will have less effect in preventing combustion, particularly of those bodies requiring low temperatures, than gases at the common heat of the atmosphere.

I have made some experiments on the effects of steam, and their results were conformable to these views. I found that a

very large quantity of steam was necessary
to prevent sulphur from burning. Oxygene
and hydrogene exploded by the electric
spark when mixed with five times their volume of steam; and even a mixture of air
and carburetted hydrogene gas, the least
explosive of all mixtures, required a third
of steam to prevent its explosion, whereas
one-fifth of azote produced the effect.—
These trials were made over mercury, heat
was applied to water above the mercury, and
37.5 for 100 parts was regarded as the cor-
rection for the expansion of the gases.

It is probable that with certain heated
mixtures of gases, where the non-support-
ing or non-inflammable elastic fluids are in
great quantities, combination with oxygene
will take place, as in the instance mentioned, page 83, of hydrogene and chlorine, without any light, for the temperature produced will not be sufficient to render elastic media luminous; and there are
no combustions, except those of the compounds of phosphorus and the metals, in
which solid matters are the result of combinations with oxygene. I have shewn, page
53, that the light of common flames depends
almost entirely upon the deposition, ignition

and combustion of solid charcoal; but to produce this deposition from gaseous substances demands a high temperature. Phosphorus, which rises in vapour at common temperatures, and the vapour of which combines with oxygen at those temperatures, as I have mentioned before, is always luminous, for each particle of acid formed must, there is every reason to believe, be white hot; but so few of these particles exist in a given space that they scarcely raise the temperature of a solid body exposed to them, though, as in the rapid combustion of phosphorus, where immense numbers are existing in a small space, they produce a most intense heat.

In all cases the quantity of heat communicated by combustion, will be in proportion to the quantity of burning matter coming in contact with the body to be heated. Thus, the blow-pipe and currents of air operate. In the atmosphere, the effect is impeded by the mixture of azote, though still it is very great: with pure oxygene compression produces an immense effect, and with currents of oxygene and hydrogene, there is every reason to believe, that solid matters are made to attain the temperature of the flame. This temperature, however, evidently presents the limit to experiments of this kind, for bodies exposed to flame can never be hotter than flame itself; whereas in the voltaic apparatus there seems to be no limit to the heat, except the volatilization of the conductors.

The temperatures of flames are probably very different. Where, in chemical changes, there is no change of volume, as in the instance of the mutual action of chlorine and hydrogene, prussic gas (cyanogen) and oxygene, approximations to their temperatures may be gained from the expansion in explosion.

I have made some experiments of this kind by detonating the gases by the electrical spark in a curved tube containing mercury or water; and I judged of the expansion from the quantity of fluid thrown out of the tube: the resistance opposed by mercury, and its great cooling powers, rendered the results very unsatisfactory in the cases in which it was used; but with water, cyanogen and oxygene being employed, they were more conclusive. Cyanogen and oxygene, in the proportion of one to two,

detonated in a tube of about two-fifths of
an inch in diameter, displaced a quantity
of water which demonstrated an expansion
of fifteen times their original bulk. This
would indicate a temperature of above 5000°
of Fahrenheit, and the real temperature is
probably much higher; for heat must be
lost by communication to the tube and the
water. The heat of the gaseous carbon in
combustion in this gas, appears more intense than that of hydrogene; for I found that
a filament of platinum was fused by a flame
of cyanogen in the air which was not fused
by a similar flame of hydrogene.

4. Some general observations, and practical inferences.

The knowledge of the cooling power of
elastic media in preventing the explosion of
the fire-damp, led me to those practical researches which terminated in the discovery
of the wire-gauze safe-lamp; and the general investigation of the relation and extent of these powers, serves to elucidate the
operation of wire-gauze and other tissues
or systems of apertures permeable to light
and air, in intercepting flame, and confirms

the views I originally gave of the phenomenon.

Flame is gaseous matter heated so highly
as to be luminous, and that to a degree of
temperature beyond the white heat of solid
bodies, as is shewn by the circumstance,
that air not luminous will communicate this
degree of heat.* When an attempt is
made to pass flame through a very fine
mesh of wire-gauze at the common temperature, the gauze cools each portion of the
elastic matter that passes through it, so as
to reduce its temperature below that degree
at which it is luminous, and the diminution
of temperature must be proportional to the
smallness of the mesh and the mass of the
metal. The power of a metallic or other
tissue to prevent explosion, will depend upon
the heat required to produce the combustion
as compared with that acquired by the tissue;
and the flame of the most inflammable sub-

* This is proved by the simple experiment of holding
a fine wire of platinum about the one-twentieth of an
inch from the exterior of the middle of the flame of a
spirit lamp, and concealing the flame by an opaque
body, the wire will become white hot in a space where
there is no visible light.

stances, and of those that produce most heat
in combustion, will pass through a metallic
tissue that will interrupt the flame of less
inflammable substances, or those that produce
little heat in combustion. Or the tissue being
the same, and impermeable to all flames at
common temperatures, the flames of the
most combustible substances, and of those
which produce most heat, will most readily
pass through it when it is heated, and each
will pass through it at a different degree of
temperature.   In short, all the circumstances which apply to the effect of cooling
mixtures upon flame, will apply to cooling
perforated surfaces.   Thus, the flame of
phosphuretted hydrogene at common temperatures, will pass through a tissue sufficiently large not to be immediately choaked
up by the phosphoric acid formed, and the
phosphorus deposited.*   A tissue of 100
apertures to the square inch, made of wire
of one-sixtieth, will at common temperatures

* If a tissue containing above 700 apertures to the
square inch be held over the flame of phosphorus or
phosphuretted hydrogene, it does not transmit the flame
till it is sufficiently heated to enable the phosphorus to
pass through it in vapour. Phosphuretted hydrogene is
decomposed in flame, and acts exactly like phosphorus.

intercept the flame of a spirit lamp, but not that of hydrogene; and when strongly heated, it will no longer arrest the flame of the spirit lamp. A tissue which will not interrupt the flame of hydrogene when red hot, will still intercept that of olefiant gas, and a heated tissue which would communicate explosion from a mixture of olefiant gas and air, will stop an explosion from a mixture of fire-damp, or carburetted hydrogene.

The ratio of the combustibility of the different gaseous matters are likewise to a certain extent as the masses of heated matter required to inflame them.* Thus an iron wire of one-fortieth of an inch heated cherry red, will not inflame olefiant gas, but it will inflame hydrogene gas; and a wire of one-eighth, heated to the same degree, will inflame olefiant gas; but a wire of one-five-hundredth must be heated to whiteness to inflame hydrogene, though

* It appeared to me in these experiments, that the worst conducting and best radiating substances required to be heated higher for equal masses to produce the same effect upon the gases; thus red hot charcoal had evidently less power of inflammation than red hot iron.

at a low red heat it will inflame bi-phos-phuretted hydrogene gas ; but wire of one-fortieth heated even to whiteness will not inflame mixtures of fire-damp.

These circumstances will explain, why a mesh of wire so much finer is required to prevent the explosion from hydrogene and oxygene from passing, and why so coarse a texture and wire is sufficient to prevent the explosion of the fire-damp, fortunately the least combustible of the known inflammable gases.

The general doctrine of the operation of wire gauze cannot be better elucidated than in its effects upon the flame of sulphur. When wire gauze of 600 or 700 apertures to the square inch is held over the flame, fumes of condensed sulphur immediately come through it, and the flame is inter-cepted; the fumes continue for some in-stants, but as the heat increases they di-minish, and at the moment they disappear, which is long before the gauze becomes red hot, the flame passes; the temperature at which sulphur burns being that at which it is gaseous.

Another very simple illustration of the

truth of this view is offered in the effect of the cooling agency of metallic surfaces upon very small flames. Let the smallest possible flame be made by a single thread of cotton immersed in oil, and burning immediately upon the surface of the oil: it will be found to be about one-thirtieth of an inch in diameter. Let a fine iron wire of one-one-hundred-and-eightieth be made into a circle of one-tenth of an inch in diameter and brought over the flame. Though at such a distance, it will instantly extinguish the flame, if it be *cold*: but if it be held above the flame, so as to be slightly heated, the flame may be passed through it. That the effect depends entirely upon the power of the metal to abstract the heat of flame, is shown by bringing a glass capillary ring of *the same* diameter and size over the flame; this being a much worse conductor of heat, will not extinguish it even when *cold*. If its size, however, be made greater, and its circumference smaller, it will act like the metallic wire, and require to be heated to prevent it from extinguishing the flame.*

* Let a small globe of metal of one-twentieth of an

Suppose a flame divided by the wire gauze into smaller flames, each flame must be extinguished in passing its aperture till that aperture has attained a temperature sufficient to produce the permanent combustion of the explosive mixture.

A flame of sulphur may be made much smaller than that of hydrogene, that of hydrogene smaller than that of a wick fed with oil, and that of a wick fed with oil smaller than that of carburetted hydrogene; and a ring of cool wire which instantly extinguishes the flame of carburetted hydrogene, only slightly diminishes the size of a flame of sulphur of the same dimensions.

Where rapid currents of explosive mixtures are made to act upon wire gauze, it is of course much more rapidly heated; and therefore the same mesh which arrests the flames of explosive mixtures at rest,

inch in diameter, made by fusing the end of a wire be brought near a flame of one-thirtieth in diameter, it will extinguish it when cold at the distance of its own diameter; let it be heated, and the distance will diminish at which it produces the extinction; and at a white heat it does not extinguish it by actual contact, though at a dull red heat it immediately produces the effect.

will suffer them to pass when in rapid motion; but by *increasing* the cooling surface by diminishing the size, or increasing the depth of the aperture, all *flames*, however rapid their motion, may be arrested. Precisely the same law applies to explosions acting in close vessels : very minute apertures when they are only few in number will permit explosions to pass, which are arrested by much larger apertures when they fill a whole surface. A small aperture was drilled at the bottom of a wire gauze lamp in the cylindrical ring which confines the wire gauze; this, though less than one-eighteenth of an inch in diameter, passed the flame and fired the external atmosphere, in consequence of the whole force of the explosion of the thin stratum of the mixture included within the cylinder driving the flame through the aperture; though, had the whole ring been composed of such apertures separated by wires, it would have been perfectly safe.

Nothing can demonstrate more decidedly than these simple facts and observations, that the interruption of flame by solid tissues permeable to light and air, depends upon no recondite or mysterious cause, but to their cooling powers, simply considered as such.

When a light included in a cage of wire gauze is introduced into an explosive atmosphere of fire-damp at rest, the maximum of heat is soon obtained, the radiating power of the wire, and the cooling effect of the atmosphere, more efficient from the mixture of inflammable air, prevents it from ever arriving at a temperature equal to that of dull redness. In rapid currents of explosive mixtures of fire-damp, which heat common gauze to a higher temperature, twilled gauze, in which the radiating surface is considerably greater, and the circulation of air less, preserves an equal temperature. Indeed the heat communicated to the wire by combustion of the fire-damp in wire gauze lamps, is completely in the power of the manufacturer, for by diminishing the apertures and increasing the mass of metal, or the radiating surface, it may be diminished to any extent.

I have lately had lamps made of thick twilled gauze of wires of one-fortieth, sixteen to the warp, and thirty to the weft, which being rivetted to the screw, cannot be displaced; from its flexibility it cannot be broken, and from its strength cannot be crushed, except by a very strong blow.

Even in the common lamps the flexibility of the material has been found of great importance, and I could quote one instance of a dreadful accident having been prevented, which must have happened had any other material than wire gauze been employed in the construction of the lamp: and how little difficulty has occurred in the practical application of the invention, is shown by the circumstance, that it has been now for ten months in the hands of hundreds of common miners in the most dangerous mines in Britain, during which time not a single accident has occurred where it has been employed, whilst in other mines, much less dangerous, where it has not yet been adopted, some lives have been lost, and many persons burned.

The facts stated in pages 74 and 75 explain why so much more heat is obtained from fuel when it is burnt quickly; and they show that in all cases the temperature of the acting bodies should be kept as high as possible, not only because the general increment of heat is greater, but likewise, because those combinations are prevented which at lower temperatures take place without any considerable production of heat: thus, in the Argand lamp, the Liverpool lamp, and in the best fire-places, the increase of effect does not depend merely upon the rapid current of air, but likewise upon the heat preserved by the arrangements of the materials of the chimney, and communicated to the matters entering into inflammation.

These facts likewise explain the methods by which temperature may be increased, and the limit to certain methods. Currents of flame, as it was stated in the last section, can never raise the heat of bodies exposed to them, higher than a certain degree, their own temperature; but by compression, there can be no doubt, the heat of flames from pure supporters and combustible matter may be greatly increased, probably in the ratio of their compression. In the blow-pipe of oxygene and hydrogene, the maximum of temperature is close to the aperture from which the gases are disengaged, i. e. where their density is greatest. Probably a degree of temperature far beyond

any that has been yet attained, may be
produced by throwing the flame from compressed oxygene and hydrogene into the
voltaic arc,* and thus combining the two
most powerful agents for increasing temperature.

The circumstances above-mentioned combined with those noticed in page 51, explain the nature of the light of flames and
their form. When in flames pure gaseous
matter is burnt, the light is extremely
feeble : the density of a common flame
is proportional to the quantity of solid
charcoal first deposited and afterwards
burnt. The form of the flame is conical,
because the greatest heat is in the centre of
the explosive mixture. In looking sted-
fastly at flame, the part where the combustible matter is volatilized is seen, and it
appears dark, contrasted with the part in
which it begins to burn, that is where it is

---

* This experiment has been tried at my request, with
the Great Voltaic Battery of the Royal Institution. The
light of solid bodies ignited by the effect of the electricity
passing through the most intense flame known, had a
brilliancy like that of the sun, and small masses of any
kind of matter placed in it, (even magnesia) were instantly fused.

so mixed with air as to become explosive. The heat diminishes towards the top of the flame, because in this part the quantity of oxygene is least. When the wick increases to a considerable size from collecting charcoal, it cools the flame by radiation, and prevents a proper quantity of air from mixing with its central part; in consequence, the charcoal thrown off from the top of the flame is only red hot, and the greater part of it escapes unconsumed.

The intensity of the light of flames in the atmosphere is increased by condensation, and diminished by rarefaction, apparently in a higher ratio than their heat, more particles capable of emitting light exist in the denser atmospheres, and yet most of these particles in becoming capable of emitting light, absorb heat; which could not be the case in the condensation of a pure supporting medium.

The facts stated in Section I. of this extract, show that the luminous appearances of shooting stars and meteors cannot be owing to any inflammation of *elastic* fluids, but must depend upon the ignition of solid bodies."

6. *Some new Experiments and Observations on the Combustion of Gaseous Mixtures,* * *&c.*

" In the last paper, I have described the phenomena of the slow combustion of hydrogene and olefiant gas without flame. In the same paper I have shown, that the temperature of flame is infinitely higher than that necessary for the ignition of solid bodies. It appeared to me, therefore, probable, that in certain combinations of gaseous bodies, for instance, those above referred to, when the increase of temperature was not sufficient to render the gaseous matters themselves luminous; yet still it might be adequate to ignite solid matters exposed to them. I had devised several experiments on this subject. I had intended to expose fine wires to oxygene and olefiant gas, and to oxygene and hydrogene during their slow combination under different circumstances, when I was accidentally led to the knowledge of the *fact,* and, at the same time, to the discovery of a new and curious series of phenomena.

I was making experiments on the increase

---

* Phil. Trans. Jan. 1817.

of the limits of the combustibility of gaseous
mixtures of coal gas and air by increase of
temperature. For this purpose, I introduced
a small wire-gauze safe-lamp with some fine
wire of platinum fixed above the flame, into
a combustible mixture containing the maximum of coal gas, and when the inflammation
had taken place in the wire-gauze cylinder,
I threw in more coal gas, expecting that the
heat acquired by the mixed gas in passing
through the wire-gauze would prevent the
excess from extinguishing the flame. The
flame continued for two or three seconds
after the coal gas was introduced; and when
it was extinguished, that part of the wire of
platinum which had been hottest remained
ignited, and continued so for many minutes,
and when it was removed into a dark room,
it was evident that there was no flame in the
cylinder.

It was immediately obvious that this was
the result which I had hoped to attain by
other methods, and that the oxygene and
coal gas in contact with the hot wire combined without flame, and yet produced heat
enough to preserve the wire ignited, and to
keep up their own combustion. I proved

the truth of this conclusion by making a
similar mixture, heating a fine wire of platinum and introducing it into the mixture.
It immediately became ignited nearly to
whiteness, as if it had been itself in actual
combustion, and continued glowing for a
long while; and when it was extinguished,
the inflammability of the mixture was found
entirely destroyed.

A temperature much below ignition only
was necessary for producing this curious
phenomenon, and the wire was repeatedly
taken out and cooled in the atmosphere till
it ceased to be visibly red; and yet when
admitted again, it instantly became red hot.

The same phenomena were produced with
mixtures of olefiant gas and air. Carbonic
oxide, prussic gas and hydrogene, and in the
last case with a rapid production of water;
and the degree of heat I found could be regulated by the thickness of the wire. The
wire, when of the same thickness, became
more ignited in hydrogene than in mixtures
of olefiant gas, and more in mixtures of olefiant gas than in those of gaseous oxide of
carbon.

When the wire was very fine, about the one-eightieth of an inch in diameter, its heat increased in very combustible mixtures, so as to explode them. The same wire in less combustible mixtures only continued bright red, or dull red, according to the nature of the mixture.

In mixtures not explosive by flame within certain limits, these curious phenomena took place whether the air or the inflammable gas was in excess.

The same circumstance occurred with certain inflammable vapours. I have tried those of ether, alcohol, oil of turpentine, and naphtha. There cannot be a better mode of illustrating the fact, than by an experiment on the vapour of ether or of alcohol, which any person may make in a minute. Let a drop of ether be thrown into a cold glass, or a drop of alcohol into a warm one. Let a few coils of wire of platinum of the one-sixtieth or one-seventieth of an inch be heated at a hot poker or a candle, and let it be brought into the glass; it will in some part of the glass become glowing, almost white hot, and will continue so as long as a sufficient quantity of vapour and of air remain in the glass *.

When the experiment on the slow combustion of ether is made in the dark, a pale phosphorescent light is perceived above the wire, which of course is most distinct when the wire ceases to be ignited. This appearance is connected with the formation of a peculiar acrid volatile substance possessed of acid properties.

The chemical changes in general produced by slow combustion appear worthy of investigation. A wire of platinum introduced under the usual circumstances into a mixture of prussic gas, (cyanogen) and oxygene in excess became ignited to whiteness, and the yellow vapours of nitrous acid were observed in the mixture. And in a mixture of

---

* The same phenomena are produced by the vapour of camphor. An ingenious form of this experiment has been lately shown in a lamp which has been called *the lamp without flame*, and which is sold in the chemists' shops. A fine wire of platinum of the one-hundredth of an inch is placed in coils round the wick of a lamp fed with spirits of wine, and a little above it. When the flame of the lamp is blown out, the heat of the wire is sufficient to keep up the slow combustion necessary for its continued ignition.

olefiant gas non-explosive from the excess of inflammable gas, much carbonic oxide was formed.

I have tried to produce these phenomena with various metals; but I have succeeded only with platinum and palladium; with copper, silver, iron, gold, and zinc, the effect is not produced. Platinum and palladium have low conducting powers, and small capacities for heat compared with other metals, and these seem to be the principal causes of their producing, continuing, and rendering sensible these slow combustions.

I have tried some earthy substances which are bad conductors of heat; but their capacities and power of radiating heat appear to interfere. A thin film of carbonaceous matter entirely destroys the igniting power of platinum, and a slight coating of sulphuret deprives palladium of this property, which must principally depend upon their increasing the power of the metals to radiate heat.

Thin laminæ of the metals, if their form admits of a free circulation of air, answer as well as fine wires; and a large surface of platinum may be made red hot in the vapour of ether, or in a combustible mixture of coal gas and air.

I need not dwell upon the connection of these facts respecting slow combustion, with the other facts I have described in the history of flame. Many theoretical views will arise from this connection, and hints for new researches, which I hope to be able to pursue. I shall now conclude by a practicable application. By hanging some coils of fine wire of platinum, or a fine sheet of platinum or palladium above the wick of his lamp, in the wire gauze cylinder, the coal miner, there is every reason to believe, will be supplied with light in mixtures of fire-damp no longer explosive; and should his flame be extinguished by the quantity of fire-damp, the glow of the metal will continue to guide him, and by placing the lamp in different parts of the gallery, the relative brightness of the wire will show the state of the atmosphere in these parts. Nor can there be any danger with respect to respiration whenever the wire continues ignited, for even this phenomenon ceases when the foul air forms about two-fifths of the volume of the atmosphere.

I introduced into a wire gauze safe-lamp a small cage made of fine wire of platinum of the one-seventieth of an inch in thickness, and fixed it by means of a thick wire of platinum about two inches above the wick which was lighted. I placed the whole apparatus in a large receiver, in which, by means of a gas holder, the air could be contaminated to any extent with coal gas. As soon as there was a slight admixture of coal gas, the platinum became ignited; the ignition continued to increase till the flame of the wick was extinguished, and till the whole cylinder became filled with flame; it then diminished. When the quantity of coal gas was increased so as to extinguish the flame; at the moment of the extinction the cage of platinum became white hot, and presented a most brilliant light. By increasing the quantity of the coal gas still farther, the ignition of the platinum became less vivid. When its light was barely sensible, small quantities of air were admitted, its heat speedily increased; and by regulating the admission of coal gas and air it again became white hot, and soon after lighted the flame in the cylinder, which as usual, by the addition of more atmospherical air, re-kindled the flame of the wick.

This experiment has been very often repeated, and always with the same results· When the wire for the support of the cage' whether of platinum, silver, or copper, was very thick, it retained sufficient heat to enable the fine platinum wire to re-kindle in a proper mixture a half a minute after its light had been entirely destroyed by an atmosphere of pure coal gas; and by increasing its thickness the period might be made still longer.

The phenomenon of the ignition of the platinum takes place feebly in a mixture consisting of two of air and one of coal gas, and brilliantly in a mixture consisting of three of air and one of coal gas: the greater the quantity of heat produced the greater may be the quantity of the coal gas, so that a large tissue of wire will burn in a more inflammable mixture than single filaments, and a wire made white hot will burn in a more inflammable mixture than one made red hot. If a mixture of three parts of air and one of fire-damp be introduced into a bottle, and inflamed at its point of contact with the atmosphere, it will not explode, but will burn like a pure inflammable substance. If a fine wire of platinum coiled at its end be slowly passed through the flame, it will continue ignited in the body of the mixture, and the same gaseous matter will be found to be inflammable and to support combustion.

There is every reason to hope that the same phenomena will occur with the cage of platinum in the fire-damp, as those which have been described in its operation on mixtures of coal gas. In trying experiments in fire-damp, *the greatest care must be taken that no filament or wire of platinum protrudes on the exterior of the lamp, for this would fire externally an explosive mixture.* However small the mass of platinum which kindles an explosive mixture in the safe-lamp, 'the result is the same as when large masses are used ; the force of the explosion is directed to, and the flame arrested by, the whole of the perforated tissue.

When a large cage of wire of platinum is introduced into a very small safe-lamp, even

explosive mixtures of fire-damp are burnt
without flame ; and by placing any cage of
platinum in the bottom of the lamp round
the wick, the wire is prevented from being
smoked."

7. Explanation of the Plate representing different Forms of the Miners' Safety Lamp, and of the Apparatus for giving Light in explosive Mixtures.

" *a.* Represents the single cylinder of wire gauze; the foldings *α. α. α.* must be very well doubled and fastened by wire. If the cylinder be of twilled wire gauze, the wire should be at least of the thickness of one-fortieth of an inch, and of iron or copper, and 30 in the warp and 16 or 18 in the weft. If of plain wire gauze, the wire should not be less than one-sixtieth of an inch in thickness, and from 28 to 30 both warp and woof.

b. represents the second top which fits upon *a.*

c. represents a cylinder of brass, in which the wire gauze is fastened by a screw to prevent it from being separated from the lamp by any blow. *c.* is fitted into a female screw, which receives the male screw *β,* of the lamp *f. f.* is the lamp furnished with its safe trimmer and safe feeder for oil.

A. is the wire gauze lamp put together with its strong wire supports, which may be three or four receiving the handle.

J. is a small cage made of wire of platinum, of one-seventieth or one-eightieth of an inch in thickness, fastened to a wire for raising it above the wick, for giving light in inflammable media, containing too little air to be explosive.

h. is a similar cage for placing in the bottom of the lamp, to prevent it from being smoked by the wick.

C. is a lamp of which the cylinder is copper of one-fortieth of an inch in thickness, perforated with longitudinal apertures of not more than the one-sixteenth of an inch in length, and the one-thirtieth in breadth. In proportion as the copper is thicker, the apertures may be increased in size. This form of a lamp may be proper where such an instrument is only to be occasionally used, but for the general purposes of the collier, wire gauze, from its flexibility, and the ease with which new cylinders are introduced, is much superior *.

D. is a lamp fitted with a tin-plate mirror

* In the first lamps which I made on this plan, the apertures were circular ; but in this case their diameters were required to be very small, as the circular aperture is the most favourable to the transmission of flame.

of half the circumference of the cylinder, and reaching as high as the single top, which may be used in strong currents of fire-damp to prevent the heat from rising too high.

All these forms of the wire gauze lamp are equally safe. In the twilled gauze lamp less fire-damp is burnt, and the radiating and cooling surface is greater, and it is therefore fitted for very explosive mixtures, or for explosive currents. The wire gauze lamp with a double cylinder, or with a reflector, answers the same purpose.

The general principle is, that the cylinder should in no case be suffered to be heated above dull redness; and this is always effected by increasing the cooling surfaces, or by diminishing the circulation of the air."

Mr. Newman has applied a lens to the exterior of some of the lamps; which when a strong light is required to be thrown upon particular objects, or parts of the mine, has been found useful.

III. *Some extracts from communications on the application of the Safety Lamp.*

THE evidence of the use of a practical discovery is of most value when furnished by practical men. I shall therefore annex extracts from some communications on the application of the Safety Lamp. Those from Newcastle and Whitehaven, will, I am sure, derive importance from the names attached to them. That from Wales is given as affording an instance in which the combustion of the fire-damp within the lamp was sufficient to destroy it for a considerable time in the workings of a colliery, and it is amusing from the simplicity of the detail.

*I. Extract from a Letter on the practical appli-
cation of the Wire-gauze Safe-lamp, from* JOHN
BUDDLE, *Esq. to* SIR H. DAVY.

Wall's-end Colliery, Newcastle,
1st June, 1816.

" After having introduced your safety-
lamp into general use in all the collieries
under my direction, where inflammable air
prevails; and after using them daily in
every variety of explosive mixture for up-
wards of three months, I feel the highest
possible gratification in stating to you, that
they have answered to my entire satisfaction.

The safety of the lamps is so easily
proved, by taking them into any part of a
mine, charged with fire-damp, and all the
explosive gradations of that dangerous element, are so easily and satisfactorily ascertained by their application, as to strike the
minds of the most prejudiced with the
strongest conviction of their high utility;
and our colliers have adopted them with
the greatest eagerness.

In the practical application of the lamps,
scarcely any difficulty has occurred. Those

of the ordinary working size, when prepared
with common cotton wick and the Green-
land whale oil, burn during the collier's
shift, or day's work of six hours, without
requiring to be replenished; and the safety
trimmer answers the purpose of cleaning,
raising, and lowering the wick completely."

" The only inconvenience experienced
arises from the great quantity of dust, produced in some situations by working the
coal, closing up the meshes of the wire-
gauze, and obscuring the light; but the
workmen very soon removed this inconvenience by the application of a small brush.

Our colliers have found it most convenient to hang the stationary lamps from
small wooden pedestals; but on observing,
that where the side of the lamps have been
suffered to come in contact with the pedestals, the wood is charred to a considerable
depth by the heat of the lamps; I have
thought it right to use small iron pedestals
instead of the wooden ones.

Beside the facilities afforded by this
invention to the working of coal mines,
abounding in fire-damp, it has enabled
the directors and superintendants to as-

certain with the utmost precision and expedition, both the presence, the quantity, and the correct situation of the gas. Instead of creeping inch by inch with a candle, as is usual, along the galleries of a mine suspected to contain fire-damp, in order to ascertain its presence, we walk firmly on with the safe lamps, and with the utmost confidence prove the actual state of the mine. By observing attentively the several appearances upon the flame of the lamp, in an examination of this kind, the cause of accidents which have happened to the most experienced and cautious miners is completely developed; and this has hitherto been, in a great measure, matter of mere conjecture.

When the discharge of inflammable air is regular, and the density of the atmosphere continues uniform, the firing point may be judged of, and approached with safety by a common candle. But when the discharge of inflammable air is irregular, or the atmosphere is in an unsettled state, a degree of uncertainty and danger attends the experiment of ascertaining the state of a mine. With the safe-lamp, however, it is reduced

to the utmost certainty, the actual presence
and position of the gas is not only ascertained with the greatest precision, but also
every alteration of circumstance or position
is distinctly perceived.

By placing a lamp near the spot where a
quantity of inflammable air is issuing, and
mixing with the circulating current of atmospherical air to the firing point, it will be
seen that very remote causes frequently
produce pulsations in the atmosphere of the
mine, which occasion the gas to fire at
naked lights ; thus showing clearly the in-
stability of the element with which we have
to deal, and the reason why so many explosions have occurred where lights have not
approached the place where the gas was
lodged within a considerable distance.

Objections have been made by some who
have not had experience of the lamps, to
the delicacy of the wire-gauze, under the
apprehension that it may be very soon im-
paired by the flame within the cylinder.
Of this, however, I have no reason to com-
plain, as, after three months constant use,
the gauze has not, in the hands of careful
workmen, been perceptibly injured by the

action of the flame; but the outer top
gauze of one or two of Newman's making
has been worn through by the friction of
the rivet* on the bottom of the swivel, to
which the finger ring is fastened; but this
only happened to the lamps used by the
wastemen, whose business it is to travel daily
in the various avenues of the mines, for the
purpose of keeping the passage for the current of air free from obstructions: nothing
of the kind has happened to the stationary
lamps used by the colliers. In short, I do
not apprehend that the gauze can be injured
by any ordinary cause without being observed in time sufficient to prevent accidents; and that we have no danger to apprehend, except from the gross negligence
of some heedless individual, or an accident
of a very unusual description, occurring to
injure the gauze.

I find that I have extended my letter to
a greater length than I intended; but I
trust, Sir, that you will excuse me for having gone so much into detail, as I feel peculiar satisfaction in dwelling upon a sub-

* This rivet is now fixed. H. D.

ject which is of the utmost importance, not only to the great cause of humanity, and to the mining interest of this country, but also to the commercial and manufacturing interests of the United Kingdom: for I am convinced that by the happy invention of the safe-lamp, large proportions of the coal mines of the empire will be rendered available, which otherwise might have remained inaccessible—at least without an invention of similar utility, it could not have been wrought without much loss of the mineral, and risk of life and capital.

It is not necessary that I should enlarge upon the national advantages which must necessarily result from an invention calculated to prolong our supply of mineral coal, because I think them obvious to every reflecting mind ; but I cannot conclude without expressing my highest sentiments of admiration for those talents which have developed the properties and controlled the power of one of the most dangerous elements which human enterprize has hitherto had to encounter."

2. *Extracts from papers written by* John Buddle,
Esq. on the Use of the Wire Gauze Safety Lamp.

" Having observed in some of the pe-
riodical publications, certain remarks on Sir
H. Davy's lamp, which in my mind appear
to have originated in motives unconnected
with truth and the improvement of science,
I feel myself called upon to do an act of
justice to the merit of the invention, in a
public statement of its great utility and extensive use in the coal mines of this country.

During the last ten months it has been
extensively employed in all the collieries
under my inspection ; and it gives me the
highest pleasure to be able to state, that
during that time not the slightest accident
by fire has occurred from its use, though
several hundreds of lamps are daily employed.

In the parts of mines where fire-damp
prevails, the surveys and inspections are
now carried on by the light of the lamp
without apprehension of danger from explosion ; for experience has shewn us, that,
with the caution of keeping it in proper re-

pair, it is absolutely safe ; and for the truth
of this, I appeal to all my professional
brethren who have had occasion to use it,
without fear of contradiction.

The colliers never hesitate a moment to
take it into any respirable part of a mine,
however much it may be charged with fire-
damp; for, whenever it appears that the
air, either from discharges of gas, or from
casual interruptions of the circulating current, becomes explosive, only give the collier his *Davy*, (the name applied in our mines
to the safe lamp,) and he goes to his occu-
pation with the same confidence in this impure atmosphere, that he would do in any
other situation, with a candle.

There has been much quibbling about
the *perfect* safety of the wire gauze lamp. I
scarcely know how the words perfect safety
can apply to any invention for the preservation of human life; but when we have seen
some hundreds of the wire gauze lamps in
daily use for several months past, in all varieties of explosive mixture, in the most dangerous mines of this country, without the
slightest accident occurring, it seems only
reasonable to infer, that they approximate
as nearly to perfect safety as any thing of

human contrivance or manufacture can be expected to do.

It would, however, be quite unreasonable to expect that accidents are never to happen, where the wire gauze lamps are used ; for it must always be remembered, that setting aside the chance of their being damaged by some of the casualties incidental to coal mining, they are to be entrusted to the management of a body of men amongst whom negligent individuals will be found, who may use damaged lamps, or expose the naked flame to the fire damp, in spite of the utmost vigilance of the overmen and inspectors of the mines. Instances of great negligence have occurred, fortunately without any ill consequences—always with the dismissal of the offender from his employment; but it would be absurd to condemn the lamp, or even to quibble upon its want of safety, on this account."

" In the most extensive coal mines* in the north of England where the fire-damp prevails to such an extent, as to require the general use of the safety lamp, it has been

* 2d Extract, dated May, 1818.

found most advantageous, to employ a steady person to take charge of the lamps, and who is made responsible for keeping them in good order. A chamber is allotted to him, in which he keeps a number of spare lamps, together with oil and cotton for replenishing the lamps which are in use.

The brass collar of the wire gauze cylinders are secured to the bottoms of the lamps by locks, which can only be opened by the lamp keeper, so that the workmen cannot either by accident or carelessness expose themselves to danger by separating the wire gauze cylinders from the bottoms of the lamps.

After finishing their day's work, the colliers bring their *Davys* to the lamp-keeper's *cabin*, who unlocking them, takes the bottoms into his own possession and allows the colliers to take the wire gauze cylinders home for the purpose of cleaning them thoroughly.

When the colliers return to their work the following morning, the lamp keeper having replenished the lamps with oil and cotton, lights them and screws on their tops, and then examines them with the utmost care, before he delivers them for use ; but if the least injury or defect appears in the gauze, or any other part of a lamp, it is immediately set aside to be repaired, and the person to whom it belongs is supplied with a perfect one.

After having dispatched the business of the morning, the lamp keeper occupies himself during the day in walking leisurely through amongst the workmen, carrying some spare lamps with him, to replace such as may happen to be extinguished, &c.

After a little practice, the lamp keepers acquire great dexterity in the trimming, &c. of the lamps, and quickly discover the slightest defect or injury in the wire gauze.

It is scarcely necessary to observe, that the lamp keeper's cabin is always placed in a secure part of the mine, as near the workings as circumstances will permit."   J. B.

Extract of a Letter from Mr. Peile *to Sir* H. Davy.

Colliery Office, Whitehaven,
*6 th July*1 816.

" I take the liberty of adding a further

statement on your invaluable safe lamps, in the Whitehaven collieries belonging to the Earl of Lonsdale, since the first application of them in February last.

With us, the general use of the lamps in consequence of the good state of our ventilation is confined to leading workings, or trial drifts; and in two of these, lately going on in one of the pits unusually infected with fire-damp, and which previously were lighted by means of steel mills, we applied the lamps with great confidence and security.

In May last in these drifts an extraordinary discharge of fire-damp burst from the pavement of the mine, and the ventilation being at that time unavoidably obstructed, the atmosphere became so charged with fire-damp as to be nearly throughout an explosive mixture. In this situation we derived the unspeakable benefit of light from the lamps, and, notwithstanding the explosive state of the mixture, with the most perfect safety.

In several other places in the collieries the lamps are used with the same confidence: yet the discharge of fire-damp being

K

moderate, they are not much exposed to explosive mixtures.

In all the workings shewing the least appearance of fire-damp, the miners are supplied with lamps, and are particularly cautioned to use them on first entering when beginning to work, where, being satisfied of security, they occasionally resort to candles afterwards. This application of the lamp alone, is of the greatest utility, and prevents many slight explosions, and the miners from being burned ; besides superseding the necessity of depending on the judgment or discrimination of any individual to prove the existence of the fire-damp, as in the old method, by the candle flame.

From the repeated proofs made with the lamps, we cannot too strongly express our confidence in their security.

By experiment, a pint of oil, value sixpence, will about supply a lamp for six days, the ordinary time of a man's working, so that they are cheaper than candles.

If my humble testimony can in any degree promote the speedy use of the lamp in other places, it will give me great pleasure."

J. PEILE.

Extract of a Letter from Mr. John Morris, *Plas Issa, 27th Jan.* 1817. *To* John Simmons, *Esq. Paddington-house.*

" Sir;—You will be pleased to recollect that some time in the month of June last, I applied to you with a request you would send me immediately some of Sir Humphry Davy's safety lamps, in consequence of an explosion of the fire-damp taking place in one of your coal mines, by which several of the men were dreadfully burnt and bruised. On the arrival of the safety lamps no accurate account of their use accompanied them. But I at length obtained (I think) the Edinburgh Review, in which was a detail of some experiments. This I read to the colliers, which gave them some confidence in the lamps, prior to which they secretly treated them with silent contempt; and I found, notwithstanding these interesting details, that a great doubt existed in their minds. I therefore was obliged to give the most peremptory orders to prepare to descend, and assisting in every preparation and execution myself. But the men's wives, &c. had collected, and made so much noise and lamen-

K 2

tations, that it was with some difficulty I could keep them off: having got over this obstacle, and the men down in the pit, instantaneous destruction was momentarily anticipated when the least noise was heard. I, however, had not the least alarm or the smallest doubt of success, and consequently did all I could to remove their dreadful anxiety. The men had no sooner descended than the enemy was discovered, which they say very much alarmed them, and they would have retreated if they could, but finding that impossible, took courage, and soon found they had destroyed the enemy so far; advancing a little farther, they found him again, and again destroyed him, and so on through the whole work. Thus the first alarm was got over, when all the knowing men in the neighbourhood were got collected together to hear the result, all of which were astonished and amazed, that so simple looking an instrument should destroy and defy an enemy, heretofore unconquerable. The same precaution and use of the lamp, was gone through the second day, and when the damp was destroyed, we began working and continued to work in this way for some weeks." GEO. MORRIS.

I am possessed of a great number of similar documents respecting the use and application of the lamp; but they are in general so flattering that I might well be suspected of vanity, if they were to be laid before the public. It has been said that the coal miners have been in many instances disinclined to adopt the safety lamp, and that the proprietors of coal mines have not been always disposed to urge its application. I am anxious that this should be contradicted, for I believe there are no dangerous mines in Great Britain where the safety lamp is not well known, and its efficacy properly estimated, and it is likewise getting into general use on the Continent.

It would be expecting too much from human nature to suppose that there should be no instances of obstinacy in workmen, and of prejudice or indifference in coal owners; but these instances have been solitary ones; and if delicacy did not forbid me, I might bring forward numerous proofs of the gratitude and warm feelings with which this invention has been received by the individuals who have benefited by it. I might appeal to the letters of thanks of various individuals, and of the united colliers of Whitehaven, to the vote of thanks of the coal trade of the north of England, of the grand Jury of Durham, of the Chamber of Commerce of Mons, and above all, to the present made to me at Newcastle in so flattering a manner, October 11, 1817.

IV. *Conclusion.—Some Practical Observations.*

TIME of course has increased the confidence of the workmen in the safety lamp; and in a period of nearly two years and a half in the most dangerous mines in Britain, it must have been exposed to all circumstances which the variety of explosive mixtures can occasion. In my first trials with the safety lamp when I found the wire become red hot, I thought it might be proper to cool it occasionally by water, or moistened cloth; but this was found unnecessary in the common practice of the miners. Whenever a single wire gauze lamp is made to burn in a very explosive atmosphere at rest, the heat soon arrives at its maximum, and then diminishes; and the idea of the wires burning out, is shewn to be unfounded; the carbonaceous matter produced from the decomposition of the oil, tends not only to prevent the oxidation of the metal, but likewise revives any oxide already formed; and this coaly matter, when the fire-damp is burning in the lamp, choaks the upper apertures of the cylinder, and gradually diminishes the heat, by diminishing the quantity of gas consumed.

I have seen wire gauze lamps in the hands of workmen, which they had used for several months, and which had been often red hot in explosive atmospheres, and which were still perfect.

Where an explosive mixture is in rapid motion, it produces, as has been stated page 97, much more heat: and in this case the radiating or cooling surfaces of the lamp must be increased, or the circulation of air diminished. Twilled gauze, or a double or triple-fold of wire gauze on one side of the lamp, or a screen of metal opposite to the direction of the current, or a semi-cylinder of glass or of mica within, answers perfectly the object of preventing the heat from rising to redness.

Single iron wire gauze of the kind used in the common miner's lamp, is impermeable to the flame of all currents of fire-damp, as long as it is not heated above redness; but if the iron wire be made to burn, as at a strong welding heat, of course it can be no

longer safe ; and though such a circumstance can perhaps never happen in a colliery, yet it ought to be known and guarded
against.

I had an excellent opportunity, Septem-
ber, 1816, of making experiments on a most
violent blower, at a mine belonging to J. G.
Lambton, Esq. This blower is walled off
from the mine and carried to the surface,
where it is discharged with great force. It
is made to pass through a leathern pipe, so
as to give a stream, of which the force was
felt at about two feet from the aperture in a
strong current of air. The common single
working lamps and double gauze lamps
were brought upon this current, both in the
free atmosphere and in a confined air. The
gas fired in the lamps in various trials, but
did not heat them above dull redness, and
when they were brought far into the stream
they were finally extinguished.

A brass pipe was now fixed upon the
blower tube, so as to make the whole stream
pass through an aperture of less than half
an inch in diameter, which of course formed
a most powerful blow-pipe, from which the
fire-damp, when inflamed, issued with great

violence and a roaring noise, making an intense flame of the length of five feet. The blow-pipe was exposed at right angles to a strong wind, and double gauze lamps and single lamps successively placed in it. The double gauze lamps soon became red hot at the point of action of the two currents ; but the wire did not burn, nor did it communicate explosion. The single gauze lamp did not communicate explosion, as long as it was red hot and slowly moved through the currents ; but when it was fixed at the point of most intense combustion, it reached a welding heat, the iron wire began to burn with sparks, and the explosion then passed.

In a second and third set of experiments on this violent blow-pipe of fire-damp, single lamps, with slips of tin-plate on the outside or in the inside, to prevent the free passage of the current, and double lamps, were exposed to all the circumstances of the blast, both in the open air and in an engine-house where the atmosphere was explosive to a great extent round the pipe, and through which there was a strong current of atmospheric air ; but the heat of the wire never approached near the point at which iron wire burns, and the explosion could never be communicated. The flame of the fire-damp flickered and roared in the lamps, but did not escape from its prison.

There is no reason ever to expect a stream of gas of this kind in a mine ; but, if it should occur, the mode of reaching and examining it, with most perfect security, is shown ; and the lamp offers a resource, which can never exist in a steel-mill, the sparks of which would undoubtedly inflame a current of this kind.

If a workman having only a common single lamp, finds the temperature of the wire increasing rapidly in an explosive mixture near a blower, he can easily diminish the heat by turning his back upon the current, and keeping it from playing upon the lamp, by means of his clothes or his body ; or by bringing the lamp nearer the orifice, from which the fire-damp issues, he may extinguish it ; and there never can be any occasion for him to place his lamp in the exact *point* where two currents one of fresh air and one of fire-damp meet each other.

When the fire-damp is inflamed in the

wire gauze cylinders, coal dust thrown into the lamp burns with strong flashes and scintillations; the miners were at first alarmed by an effect of this kind produced by the dust naturally raised during the working of the coals. I have made a number of experiments on this subject; but though I have repeatedly thrown coal dust, powdered rosin, and witch meal, through lamps burning in more explosive mixtures than ever occur in coal mines, and though I have kept these substances floating in the explosive atmospheres, and heaped them upon the top of the lamp when it was red hot, yet I never could communicate explosion by means of them. Phosphorus or sulphur are the only substances which can produce explosion by being applied to the outside of the lamp; and sulphur, to produce the effect, must be applied in large quantities, and blown upon by a current of fresh air.

It will be unnecessary to caution the workmen against heaping sulphur, or gunpowder, or pyrites, which afford sulphur by distillation, upon their lamps; and such dust from these substances as can float in the atmosphere cannot produce inflammation; for minute particles of ignited solid matter have no power of inflaming the fire-damp; and I have repeatedly blown fine coal dust mixed with minute quantities of the finest dust of gunpowder through the lamp burning in explosive mixtures without any communication of explosion.

A few complaints have been made of the feebleness of the light of lamps after they have been some time used, in consequence of the tissue being choaked up by coal dust. But by means of a brush this evil may be removed. And in some experiments, that I made with Mr. Buddle, in the Wall's-end Colliery, the light of a single gauze lamp furnished with a tin-plate reflector, was found to be superior to that of a common miner's candle, and the light of a lamp without the reflector, nearly equal to it; and almost double to that of the steel-mill at its greatest intensity of light. The trials were conducted by determining the distance at which an object was visible with the different species of light, and considering the intensity of the light directly as the square of the distance.

I have often made lamps in which surfaces of glass were used for transmitting
light without a guard of wire-gauze, they
give more light, but they are highly dangerous, and ought never to be used. Thick
plates of mica,* (Muscovy glass,) may, however, be safely employed, though great
care must be taken in this case that the
radiating and cooling surfaces, where the
fire-damp burns, are extensive, and all the
precautions mentioned, page 19, must be
adopted.

Where a lamp is permanently to be fixed
in a part of a mine ; it will afford a better
light if some of the surfaces are of mica ;
but for lamps which are constantly to be

* From a very able report on the safety-lamp, drawn
up by order of the Chamber of Commerce of Mons,
and furnished with some very intelligent and scientific
notes by M. Gossart, President of the Chamber, it
appears that lamps with plates of mica have been used
in Flanders. M. Gossart quotes an instance which
proves the danger of glass. " A director of the works
having descended in the colliery of Tapatouts, with a
lamp of which the base of the cylinder was of glass, a
drop of water fell upon and broke the glass, and de-
tached a piece which would have opened a communication for explosion ; but the air fortunately at the moment was not adulterated with fire-damp."

carried about by the miners, iron wire-gauze I have no doubt, will be the material always employed.—I have tried a lamp on the plan of Argand's, in which wire-gauze feeders were below, and in which a current of air was determined by wire-gauze cylinders above; it gave a brilliant light, but produced so much heat as to boil the oil in the reservoir, and it required a complicated contrivance for trimming it.

When a cage of wire of platinum is used within a lamp, even where the explosive mixture burns with flame, it diminishes the heat by its radiation, and it increases the light, so that it will always be useful in lamps; and as it is an imperishable metal, it is only the original expence, which is very small, that is to be attended to. It is proper to urge again what has been mentioned, page 112, that no wire or filament of platinum must be suffered to project beyond the wire-gauze, so as to be in the external atmosphere.

The forms of lamps may be infinitely varied; but the most convenient size for a common working lamp is from eight to ten inches in height, and two to two and a half in breadth. If the wire-gauze cylinders are larger, there is too much heat produced in them by the combustion of explosive mixtures.

In gas manufactories, spirit warehouses, or druggists' laboratories where ether is distilled, the common safety lamp may be advantageously used.—A simple mode of exhibiting its power is furnished by throwing a little ether into the bottom of a large jar; the vapour of the ether, mixing with the air, will produce an explosive atmosphere, which will burn within the wire-gauze without inflaming the ether in the bottom of the jar.

If pure hydrogene should be disengaged in any mines, the improbability of which however is very great, wire-gauze of a finer texture must be employed.—This material indeed, it is likely, will be found to possess many new applications, not only in guarding against the communication of flame, but likewise in exerting cooling agencies, wherever elastic media can be exposed to it, so as to have their temperatures lowered by its radiation.

I shall now conclude. Whatever may

be the fate of the speculative part of this enquiry, I have no anxiety as to the practical results, or as to the unimpassioned and permanent judgment of the public on the manner in which they have been developed and communicated ; and no fear that an invention for the preservation of human life and the diminution of human misery will be neglected or forgotten by posterity.

When the duties of men coincide with their interests, they are usually performed with alacrity ; the progress of civilization ensures the existence of all real improvements ; and however high the gratification of possessing the good opinion of society, there is a still more exalted pleasure in the consciousness of having laboured to be useful.

APPENDIX.

No. 1.

SINCE the researches upon Flame contained in the foregoing pages have been published, M. Gay Lussac has put into my hands a paper written some years ago by M. de Humboldt and himself, which contains some very interesting results that may be adduced as confirmations of my principles, on the causes of combustion and explosion.

MM. Gay Lussac and de Humboldt have shewn that when oxygene and hydrogene are mixed in proportions in which they cannot be fired by the electrical spark, they may be still made to combine in the proportions which can form water, by artificially raising their temperature.

MM. Gay Lussac and de Humboldt suppose that the action of electricity in producing combination is owing to the heat it produces by the compression of the elastic medium through which it passes. This idea is very ingenious, but the phenomena of decomposition by electricity, shew that there is some relation between the primary attractive powers of the chemical elements and their electrical energies.

When the common electrical or voltaic electrical spark is taken in rare air, the light is considerably diminished. I made some experiments to ascertain whether the heat was likewise diminished, and I found that this was certainly the case. Yet in a receiver that contained air sixty times rarer than that of the atmosphere, a piece of wire of platinum, placed in the centre of the luminous arc, produced by the great voltaic apparatus of the Royal Institution, became white hot; and that this was not owing to the electrical conducting powers of the platinum, was proved by repeating the experiment with a filament of glass, which instantly fused in the same position.

It is evident from this, that electrical light and heat may appear in atmospheres in which the flame of combustible bodies could not exist, and the fact is interesting from its possible application in explaining the phenomena of the Aurora Borealis and Australis.

No. 2.

M. Sementini, professor of chemistry at Naples, presented me in 1819 with a lamp, in which alcohol burnt without flame, by means of fine coils of silver wire, and afforded phenomena exactly of the same kind, as the lamp furnished with wire of platinum.

When I first discovered the phenomena of the ignition of thin filaments, of platinum and paladium, I ascertained that the temperature required for this result was much below ignition; but I did not determine the precise degree on Fahrenheit's scale. It was evident, however, from the principles laid down at page 94, that it must be lower for hydrogene than most other inflammable gases ; and lower in proportion as the wire or foil was finer.

M. Dobreiner has lately made the discovery that the finely divided and spongy platinum, obtained by precipitation from solution, becomes ignited, even at common temperatures ; and MM. Thenard, and Dulong, and other chemists, in repeating his observation, have found that various metals in a finely divided state, have the same property of hastening or producing combinations at a lower temperature, than those at which they usually occur, and have given many facts analogous to those described page 107. Mr. Garden has found the air of iridium, likewise possessed of the property of inflaming mixtures of common air and hydrogene.

It is probable that the rationale of all these processes is of the same kind. Whenever any chemical operation is produced by an increase of temperature, whatever occasions an accumulation of heat, must tend to give greater facility to the process ; a very thick wire of platinum does not act upon a mixture of oxygene and hydrogene, at a heat below redness ; but if beat into thin laminæ, it occasions its combustion at the heat of boiling mercury ; and when in the form of the thinnest foil, at usual temperatures. I cooled the spongy platinum to 3° of Fah^{t.} and still it inflamed hydrogene nearly of the same temperature, issuing from a tube cooled by salt and ice.

I thought that common radiant heat or light, might be necessary to the effect ; but the cooled metal and the gases acted with the same phenomena in darkness.

It may be supposed that the spongy platinum absorbs hydrogene, or that it contains oxygene ; but neither of these hypotheses will apply to the fact that I first observed, of the ignition of fine wires in different mixtures of inflammable gases and air, at temperatures so far below ignition.

A *probable* explanation of the phenomenon, may, I think, be founded upon the Electrochemical hypothesis, which I laid before the Royal Society, in 1806 ; and which has been since adopted and explained, according to their own ideas, by different philosophers.

Supposing oxygene and hydrogene to be in the relations of negative and positive, it is necessary to effect their combination, that their electricities should be brought into equilibrium or discharged. This is done by the electrical spark or flame, which offers a conducting medium for this purpose, or by raising them to a temperature, in which they become themselves conductors. Now platinum, palladium, and iridium are bodies very slightly positive with respect to oxygene ; and though good conductors of electricity, they are bad conductors and radiators of heat, and supposing them in exceedingly small masses, they offer to the gases the conducting medium necessary for carrying off, and bringing into equilibrium their electricity without any interfering energy, and accumulate the heat produced by this equilibrium. Other metals do not possess the same union of qualities, yet most of them assist combination at lower temperatures than glass, which is a non-conductor of electricity.

That spongy platinum, even when moistened, as M. Dobreiner has very lately shewn, should facilitate the combination of oxygene and hydrogene, *may* depend upon *this peculiar* electrical property; and why foil of platinum should have its power of causing oxygene and hydrogene to combine, increased by being placed, for a short time, in nitric acid, as MM. Dulong and Thenard have shewn, may be owing to this, that the slight positive charge it acquires may, in being brought into equilibrium, be a first step in the operation; and there are analogous instances.

Fine wire of platinum, I find, when conveying currents of Electricity, as in a circuit, with zinc and sulphuric acid, or charcoal and nitromuriatic acid, has not its power of acting upon gaseous mixtures sensibly increased.

No. 3.

The general inflammable air is only disengaged in coal mines; yet the salt works of Styria, Salzburg, and Upper Austria are not exempt from accidents depending upon carbonated hydrogene gas. An explosion had happened at Aussee, in 1818, a few weeks before I visited the salt works, by which several persons were killed; and the miners received the Safety Lamp, and witnessed its operation with gratitude and surprise.

The inflammable air appeared to me, in these instances, to be derived from bituminous shist.

No. 4.

I have had some correspondence with Mr. Buddle respecting the accidents which have happened in coal mines, since the discovery of the Safety Lamp. He refers them, in all cases, to the carelessness of workmen.

I should strongly recommend double lamps, in cases where miners are obliged to work for any time in explosive mixtures, or wherever currents are expected ;— or lamps with mica, or tin-plate *within* the wire gauze to prevent too great a circulation of air, (see p. 138.) It is very easy to extinguish a lamp in which the fire-damp is burning, by sliding a tin-plate cylinder over it, or by a circle of wire gauze fitting the interior in a rim of copper and moved by the termination of the trimming wire: but it is much better, in all cases of danger, to use lamps which, *under no circumstances,* can explode. Such as these described in p. 97.

Having often trusted my life to the Safety Lamp under the most dangerous circumstances, I cannot but sometimes smile when the Public papers endeavour to invalidate its security upon the opinions or evidence of certain persons who have their own nostrums for preventing the accumulation of inflammable air in mines.

I have sometimes to read letters on the improvement of the invention, by plans : most of which are discussed in the foregoing pages ; such as using glass or mica as a part of the surface for transmitting light, using double lamps, or double lamps containing a reflecting surface to prevent explosions from currents; and I have actually seen a lamp upon the rudest model of those I first made, having thick glass above, and wire gauze below, called " the newly invented Safety Lamp."

No. 5.

For gas manufactories, or houses where gas is extensively used, I should recommend the Safety Lamp with iron wire gauze, but for the use of the Navy, those with copper wire gauze are less liable to rust. As the latest instance of a ship lost, for want of a Safety Lamp, I may mention the Kent East Indiaman, which was burnt, as I am informed by the Shipping Committee, in consequence of the inflammation of Rum, by means of a common lantern.

FINIS.

H. Bryer, Printer, Bridge-street, Blackfriars, London.